THE
GRAVITY
OF MATH

Also by the Authors

The Shape of Inner Space: String Theory and the Geometry
of the Universe's Hidden Dimensions,
by Shing-Tung Yau and Steve Nadis

A History in Sum: 150 Years of Mathematics at Harvard (1825–1975),
by Steve Nadis and Shing-Tung Yau

From the Great Wall to the Great Collider: China and
the Quest to Uncover the Inner Workings of the Universe,
by Steve Nadis and Shing-Tung Yau

The Shape of a Life: One Mathematician's Search for
the Universe's Hidden Geometry,
by Shing-Tung Yau and Steve Nadis

THE GRAVITY OF MATH

HOW GEOMETRY RULES THE UNIVERSE

STEVE NADIS and SHING-TUNG YAU

Illustrations by Mei-Heng Yueh

BASIC BOOKS

New York

Basic Books
Hachette Book Group
1290 Avenue of the Americas, New York, NY 10104
www.basicbooks.com

Printed in the United States of America

First Edition: April 2024

Published by Basic Books, an imprint of Hachette Book Group, Inc. The Basic Books name and logo is a registered trademark of the Hachette Book Group.

The Hachette Speakers Bureau provides a wide range of authors for speaking events. To find out more, go to hachettespeakersbureau.com or email HachetteSpeakers@hbgusa.com.

Basic books may be purchased in bulk for business, educational, or promotional use. For more information, please contact your local bookseller or the Hachette Book Group Special Markets Department at special.markets@hbgusa.com.

The publisher is not responsible for websites (or their content) that are not owned by the publisher.

Library of Congress Cataloging-in-Publication Data

Names: Nadis, Steven J., author. | Yau, Shing-Tung, 1949– author.
Title: The gravity of math : how geometry rules the universe / Steve Nadis and Shing-Tung Yau.
Description: First edition. | New York : Basic Books, 2024. | Includes bibliographical references.
Identifiers: LCCN 2023040942 | ISBN 9781541604292 (hardcover) | ISBN 9781541604308 (ebook)
Subjects: LCSH: Gravitation—Popular works. | Gravitation—Mathematics—Popular works. | Gravitation—History—Popular works. | General relativity (Physics)—Mathematics—Popular works. | General relativity (Physics)—History—Popular works. | Black holes (Astronomy)—Mathematics—Popular works.
Classification: LCC QC178 .N24 2024 | DDC 530.11—dc23/eng/20231130
LC record available at https://lccn.loc.gov/2023040942

ISBNs: 9781541604292 (hardcover), 9781541604308 (ebook)

LSC-C

Printing 1, 2024

To our parents,
Lorraine B. Nadis and Martin Nadis
Yeuk Lam Leung and Chen Ying Chiu

Searching for the Light

In fog and clouds, our world's true forms do hide.
There've been many tireless searches, all of them denied.

But through dawn's mist, I perceive the barest flicker of light.
A sense of harmony overtakes me, the heart and mind unite.

Flower petals unfurl, lured on by the rising sun.
Swallows glide in pairs, tracing pirouettes above.

Yet the mysteries of earth and sky always beckon, never wane.
And I, as ever, am helpless in their irresistible sway.

In homage to a fleeting insight, I take a silent bow,
girding myself for the next leg of this windy, wondrous road.

The path ahead is hard, they say, even treacherous in parts,
with difficult passes to negotiate and much altitude to be gained.

I continue on, resolute, not knowing how to stop,
anxious, and forever curious, as to what I'll see at the top.

—Shing-Tung Yau, 2023

Contents

Preface

I was a high school student in Hong Kong in the early 1960s when I first heard about Albert Einstein's general theory of relativity. Calling my introduction to the subject superficial might be understating things. The fact is, I didn't know enough mathematics at the time to begin to understand the theory, and there was no one around to teach me. Nor were there any qualified instructors at the small Hong Kong college I attended a few years later. And yet I always knew, deep in my bones, that this was a deep, rich area about which I was someday, and somehow, destined to learn.

My chance came in January 1970 during the first year of my graduate studies in mathematics at the University of California, Berkeley. I had immersed myself in geometry since arriving on campus four months earlier, but as the new year began, I started auditing some lectures on general relativity, as the theory is often called, and I was surprised to find out that the thing we call gravity—which had long been described as an attractive force—was more accurately viewed as a geometric effect, a consequence of the curving, or warping, of spacetime due to the presence of massive bodies. This came as something of a revelation to me, as I had never before realized that physics could relate so closely to geometry. In my innocence (and ignorance), I had viewed them as separate subjects.

My curiosity suddenly aroused, I started wondering whether spacetime could be curved, and gravity still manifest, even in a vacuum where no matter is present at all. Initially, I had no idea that the problem I was contemplating was equivalent to one that had been posed in 1954 by the mathematician Eugenio Calabi. Once I discovered what Calabi had done, his conjecture, which had been framed in dense mathematical language with no reference whatsoever to gravity, grabbed hold of me and occupied my attention for many years to come.

For Calabi, it was a fascinating geometric problem in its own right, apart from any possible tie-ins with general relativity, and I mostly approached it as such at the time. But I was also intrigued by the links between mathematics and physics. Indeed, from that moment of heightened curiosity, which was aroused during those lectures on general relativity, much of my career has subsequently been spent dancing along the sometimes poorly delineated boundary between these two great disciplines. I've often found this to be a fruitful realm to inhabit, as ideas from mathematics have long stimulated developments in physics, just as physics breakthroughs have repeatedly spurred progress in math. This dynamic interplay has been especially evident in the field of general relativity, and that was a big motivation behind this book.

Remarkably, the general theory of relativity, which Einstein unveiled in 1915, still encompasses almost everything we know about gravity, more than a century later. My coauthor and I wanted to highlight all the mathematics that went into the theory and underlying equations formulated by Einstein—and all the help he received from mathematicians along the way—as well as all the help mathematicians have since provided in exploring the still-unfolding ramifications of that theory. Our understanding of black holes, for example, would have been severely stunted had it not been for the many insights derived from mathematics. And yet, were it not for physicists, we might not have dreamed of these amazing things in the first place.

It has been, and continues to be, an exhilarating collaboration—one that I consider myself lucky to have been part of. And it is this marvelous (though occasionally fractious) partnership between mathematicians and physicists—and some line-straddling mathematical physicists as well—that we hope to describe and celebrate in *The Gravity of Math*.

—Shing-Tung Yau
Beijing, 2023

You might say this book started small, with just a single word. Several years ago, an editor from an academic publisher contacted my coauthor, Shing-Tung Yau, out of the blue and asked him if he had any ideas for a book. "How about a book on gravity?" Yau replied.

That was not a lot to go on, but it was more than we had in 2006 when we embarked on our first book collaboration at the behest of a New York literary agent. When we finally spoke with said agent and asked him what sort of book he had in mind, or what it might be about, he replied, "I really can't say, but I'm sure you'll come up with something great."

In this case, at least, we had a seed—a word of just seven letters to build upon, yet one that is truly gargantuan in scope. Gravity is the universe's principal architect, sculpting the cosmos on the largest scales, giving rise to everything from planets to stars to superclusters stretching billions of light-years across. Yet there's still much that we don't understand. Why, for instance, is gravity so much weaker than the other forces—thirty-six orders of magnitude weaker than electromagnetism, for instance—and why has it been so difficult to come up with a unified theory in which gravity and the other three forces (strong, weak, and electromagnetic) comfortably mesh?

It's a daunting topic to take on, especially when the central figure looming over the proceedings, Albert Einstein, has already been the focus of an estimated 1,700 books—and counting. In view of the rather voluminous existing literature, Yau and I have not set out to break new ground in terms of biographical and historical scholarship. For starters, this is _not_ a book about Einstein, per se, even though he—after an arduous ten-year effort—did succeed in crafting the theory of general relativity that stands to this day. Where we hope to make a contribution is by shining a light on the mathematical underpinnings of that theory and on the tools of mathematics that have enabled researchers to explore aspects of general relativity—in some cases going astonishingly far in the absence of, or decades in advance of, any experimental data.

It might be interesting to note that the Nobel Prize–winning physicist Steven Weinberg took pretty much the opposite approach in his classic text, _Gravitation and Cosmology_. As Weinberg wrote in 1972 on the first page of the first chapter of that book: "I have tried throughout this book to delay the introduction of geometrical objects, such as the metric...and the curvature, until the use of these objects could be motivated by considerations of physics." Our book—by putting mathematics, and especially geometry, first, front, and center—offers a different perspective, and I think a useful one, due to the fact that, in many instances, developments in the physics of general relativity and subsequent extensions of that theory were motivated by, and built upon, previously established principles of mathematics.

From my vantage point, as a non-mathematician and non-physicist, the subject matter was not easy to assimilate (and I won't dare to use the word *master* here). Not to overdramatize my personal travails, or to make a laughable comparison, but I did identify with Einstein's words regarding "the years of anxious searching in the dark, with their... alternations of confidence and exhaustion" (*Notes on the Origin of the General Theory of Relativity*, 1934). I too struggled to understand the math and physics at play well enough to figure out a way of framing the discussion contained within these pages.

Fortunately, I had expert help, not only from my coauthor, Yau, who has done some groundbreaking work of his own in mathematical general relativity, among other areas. But I also received invaluable assistance from many other mathematicians and physicists, as well as from nonscientists, and I am grateful to all of them.

Here are some of the people who contributed their time and support to this enterprise, and I apologize in advance for any omissions I might have inadvertently made: Aghil Alaee, Lars Andersson, Maureen Armstrong, Abhay Ashtekar, Ken Bernstein, Michael Bernstein, Robert Bryant, Lily Chan, Yuewen Chen, Paul Chesler, Leo Corry,

Demetrios Christodoulou, Mihalis Dafermos, Simon Donaldson, Scott Field, Felix Finster, Peter Galison, Greg Galloway, Elena Giorgi, Wei Gu, Lan-Hsuan Huang, Niky Kamran, Demetre Kazaras, Jordan Keller, Enno Kessler, Gaurav Khanna, Marcus Khuri, Sergiu Klainerman, Hari Kunduri, Sarah LaBauve, Mark Lee, Martin Lesourd, Yi Li, Irene Minder, Georgios Moschidis, James Nester, Peter Olver, Frans Pretorius, Jordan Rainone, David Rowe, Burkhard Schwab, Antoine Song, Andrew Strominger, Jérémie Szeftel, Valentino Tosatti, Henry Tye, Vijay Varma, and Robert Wald.

Mei-Heng Yueh of the National Taiwan Normal University provided a great service to us by producing the illustrations for this book—a job he handled expertly, efficiently, and speedily. I'd like to make a special shout-out to Lydia Bieri, David Garfinkle, Mu-Tao Wang, and Hung-Hsi Wu, who were extremely generous with their time and were kind enough to help me, on multiple occasions, come to grips with this challenging subject matter. They exercised extraordinary patience during our many conversations—especially in view of the fact that I can be slow on the uptake. I'm indebted to all of them, as well as to everyone else mentioned above.

Great appreciation is also owed to our editor, T. J. Kelleher (with whom we've worked splendidly in the past), the editorial assistant Kristen Kim, and the book's production editor, Shena Redmond. We're thankful for their efforts and those of many others at Basic Books—including Lara Heimert, Liz Wetzel, Katherine Robertson, Amber Hoover, Brian Distelberg, Sara Sheiner, Shivani Boodhoo, and Caitlyn Budnick—who believed in this project from the start and skillfully guided us from what was an initially rough manuscript to a polished product that we could all feel proud of. Fortunately, our manuscript was placed in the extremely capable hands of our copyeditor, Charlotte Byrnes, who smoothed over countless rough edges, expertly clarifying the text from start to finish, while providing some uniformity to my haphazard application of the strictures of American English grammar.

Finally, I'd like to thank my wife, Melissa, and my daughters, Juliet and Pauline, for always being there for me—and for graciously putting up with more discussions about gravity than the average person can tolerate. I'll offer special thanks to my parents, Lorraine and Marty, who are no longer living but nevertheless put me in a position to be able to take on ambitious (and sometimes overly ambitious) projects like this. In addition, my sister, Sue, and brother, Fred, have always been supportive—even to some of the harebrained schemes I have concocted over the years.

—Steve Nadis
Cambridge, Massachusetts, 2023

Prelude

There's More Than One Way to Slice a Cone

Sometime around the year 200 BCE, the Greek mathematician Apollonius of Perga, known to his contemporaries as the Great Geometer, set out to write everything that was then known about conic sections. Conic sections are the curves created when a plane intersects, at various angles, the surface of an infinitely long, right circular cone. If this plane is perpendicular to the cone's central axis, you get a circle. If the plane is slightly tilted, you get an ellipse. If it's tilted a bit more, you get a parabola, and tilting even farther yields a hyperbola. Euclid, Apollonius's predecessor by about a century, had already written a four-volume work called *Conics*, which built upon ideas established by the mathematician Menaechmus decades earlier. However, the eight-volume set produced by Apollonius, also called *Conics*, was much more comprehensive, containing many new ideas that he had personally formulated.

That work can be rough going. An 1896 review in the journal *Nature* of a then-new translation stated: "Formal demonstrations are given of propositions [387 altogether], which we should be apt to

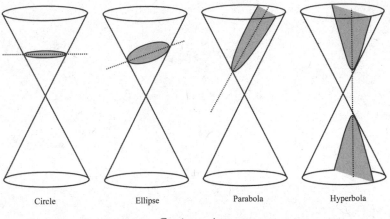

| Circle | Ellipse | Parabola | Hyperbola |

Conic sections

dismiss as intuitively evident, and a preference is shown for indirect methods of proof which, in some cases, almost amounts to perversity." Despite the long-windedness and oftentimes digressive aspects of the presentation, Apollonius's *Conics* "affords an excellent example of the methods of Greek geometry at its best period," the *Nature* reviewer added. And with one technical exception, "almost all the principal theorems of ordinary geometrical conics are to be found in this treatise, composed more than twenty centuries ago."[1]

For most of those centuries, Apollonius's work lay essentially dormant. It's often hard to tell what uses might ultimately be found for new insights in mathematics, and during that long quiescent period there'd been few, if any, practical applications nor any apparent scientific significance attached to these mathematical constructs.

All that changed at the dawn of the seventeenth century, when Johannes Kepler familiarized himself with the work of Apollonius. Kepler went on to carry out his own study of conic sections, the results of which appeared in a 1604 paper he wrote about the optical issues that can arise in astronomy. That study, in turn, sent him on a path toward a series of discoveries for which he is most famous.

In 1609, Kepler published two laws of planetary motion: The first law proclaimed that the planets of the solar system move along elliptical (rather than circular) orbits around the sun, with the sun lying at one focus of that ellipse. According to Kepler's second law, the line extending from the sun to a planet (such as Earth) sweeps out equal areas over equal periods of time. Ten years later, Kepler published his third law: The square of a planet's orbital period is proportional to the cube of its mean distance from the sun.

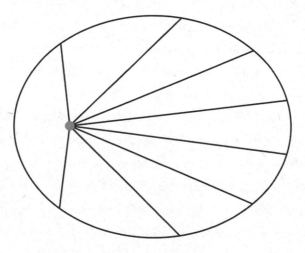

Kepler's second law

Kepler's work provided strong support for the heliocentric (as opposed to geocentric) view of the solar system, which had been advanced by Copernicus some sixty years earlier. However, Kepler was dismayed, as the physicist Robbert Dijkgraaf would later write, "that the orbits of the planets did not form the singular perfect shape of a circle, but instead had the ugly appearance of an ellipse."[2]

About sixty years later, Isaac Newton set out to explain Kepler's results. He introduced new laws of gravitation, derived from a new set of mathematical tools called calculus that he had also developed. His work showed exactly why orbiting bodies follow paths marked out by ellipses rather than circles.

For roughly 200 years, Newton's gravitational laws appeared to work flawlessly. By the end of the nineteenth century, however, some limitations had become evident. Around that time, Albert Einstein began working on a broader, more general theory of gravity, which would incorporate Newton's laws but also be equipped to handle the exceptional cases where those laws faltered.

Before he was able to make significant progress, however, Einstein had to figure out how to make use of a rather new form of geometry—or at least new to him. It had emerged some sixty years earlier, but for decades physicists had paid it little or no attention. Once Einstein was briefed on the subject by a former college classmate, however, it soon became apparent to him that this strand of mathematics might provide just what he was looking for—the foundation upon which one could build a whole new theory of gravity. And that's just what he did.

Einstein's spectacular realization, and the theory that codifies it, was described by the physicist Chen Ning Yang as "an act of pure creation" that he attributed to Einstein alone: "conceived and executed by one single person," as he put it.[3] Robbert Dijkgraaf, director of the Institute for Advanced Study where Einstein spent the last twenty-two years of his life, called the feat "perhaps the greatest achievement of a single human mind."[4] While Einstein's accomplishment surely ranks as one of the greatest theoretical breakthroughs in the history of science, it may be somewhat misleading to call it an act of "pure creation." After all, his work built on Isaac Newton's gravitational theory published more than two centuries earlier. Newton famously said, in a 1675 letter, "If I have seen further, it is by standing on the shoulders of giants." And while Einstein saw further than any of his predecessors, he also benefited from standing on the shoulders of others, especially Newton himself. Indeed, Einstein was possibly less impressed with his own work than Yang and Dijkgraaf were. Einstein called Newton a "brilliant genius who determined the course of western thought, research and practice like no one else before or since."[5]

Einstein, moreover, did not start from scratch when he began drawing up his theory of gravity; he drew upon his own special theory of relativity, the first version of which was published in 1905. And although Einstein is unquestionably the author of the general theory, it's also true that he received plenty of help along the way from others—including those whose work proved essential to him, such as Bernhard Riemann, Hermann Minkowski, and Gregorio Ricci-Curbastro. In addition, there were many who helped him personally—such as the mathematicians Marcel Grossmann, David Hilbert, and Tullio Levi-Civita—as well as Einstein's good friend, the engineer Michele Besso, whom he called "the best sounding board in Europe" for scientific ideas.[6]

It might also be worth noting that Einstein not only had to learn a lot of unfamiliar mathematics to formulate his theory, but he had previously given short shrift to that very same material. While it had never seemed worth his while to delve into these mathematical realms before, it became apparent to him that, unless he did, he would not be able to get very far. Fortunately, he did acquire the necessary mathematical skills. And far is where he did go and where—with the aid, once again, of mathematics, other mathematicians, and his own personal brilliance—his ideas have carried us ever since.

Indeed, one testament to the richness of general relativity is the fact that more than a century after its promulgation, physicists and mathematicians are still unraveling its implications and still finding new corners of the theory to push, prod, and probe. The exploration continues.

It bears repeating that this wildly successful and productive theory was built around a bit of mathematics—a form of geometry—that had been largely ignored by physicists for nearly half a century. In crafting his laws of planetary motion, Kepler similarly drew upon an 1,800-year-old treatise on conic sections, written by the "Great Geometer" of his era, which had not, until then, had much impact on physics at all. This, of course, speaks to the incredible staying power of mathematics. A rigorously proved mathematical theorem has the property—rare, if not unparalleled in human affairs—that it will hold up for all time. These axioms of eternal truth (to paraphrase Thomas Jefferson) represent tools, of sorts, that can be rediscovered later—even centuries later—and utilized in ways that its authors never dreamed of. An act of "recycling" of this very sort, which occurred in the early 1900s, lies at the heart of the story that the authors of this present book hope to tell.

= 1 =

Falling Objects, Shifting Paradigms

How does someone launch a scientific revolution? There is, to be sure, no set formula or procedures one can follow, for otherwise revolutions might be so common as to not be worthy of the name. Nevertheless, two of the greatest advances in our understanding of gravity were triggered in similar ways: the first by contemplating a randomly falling object, and the second by contemplating a person falling from the roof of a house. The notion of falling is, of course, central to our intuitive, and sometimes concrete, sense of gravity. In both cases, tremendous insights arose from speculations about free fall and the effects associated with it.

The first of the examples cited above reportedly occurred in 1666. A year earlier, in the summer of 1665, Isaac Newton had been a twenty-three-year-old fellow at Trinity College at Cambridge University. Because the Great Plague of London—a deadly outbreak of the bubonic plague—was approaching Cambridge, the university sent its students and fellows home for the better part of two years. Newton retired to his birthplace and family home in Lincolnshire, England, where he passed what proved to be an extraordinarily productive sojourn.

The following summer, while sitting in the garden and witnessing the fall of an apple, Newton "let himself be led into a deep meditation on the cause which thus draws every object along a line whose extension would pass almost through the center of the Earth," the French philosopher and author Voltaire wrote.[1] John Conduitt, who became Newton's assistant and nephew-in-law a few decades later, provided a somewhat more detailed account: While Newton was in his garden, Conduitt reported, "it came into his thought that the power of gravity (which brought an apple from a tree to the ground) was not limited to a certain distance from the Earth, but that this power must extend much further than was usually thought. Why not as high as the moon, thought he to himself, and, if so, that must influence her motion and perhaps retain her in her orbit, whereupon he fell a-calculating."[2]

Questions of this sort prompted Newton to develop a sweeping theory of gravity, which showed that the same force that draws an apple toward the center of the Earth also draws the Moon toward Earth, maintaining its regular orbit rather than drifting off into space. Several decades earlier, Johannes Kepler had suggested—based on observations and estimates rather than theoretical calculations— that planets follow elliptical paths around the sun. Assuming that Kepler had been right, and that the Moon also followed an elliptical path around Earth, Newton then showed that the Moon's trajectory would indeed be elliptical if the strength of gravity exerted by Earth on the Moon was inversely proportional to the square of the distance between them.

Newton proved his case mathematically, making use of a tool he called the theory of fluxions, and that we now call calculus, which he also began to develop during his forced hiatus at his family's country home. (Gottfried Wilhelm Leibniz independently developed calculus at roughly the same time, and a fierce priority dispute later ensued over who invented calculus. Newton's earliest efforts in this area apparently preceded those of Leibniz by about a decade, whereas Leibniz published the first papers on the subject and presented calculus in the form that was subsequently adopted by other mathematicians.[3] So it seems fair and expedient to consider both of them co-inventors of calculus and leave it at that.)

One of the new tools Newton had at his disposal, which falls under the general heading of *differential* calculus, can be utilized to determine the shape of a curve by breaking it down into infinitesimally small increments consisting of tiny line segments.

Newton also developed his own version of *integral* calculus, which can be used, for instance, to determine the area under an arbitrary curve by breaking that space down into tiny, infinitesimal rectangles. Both differential and integral calculus can be used to show that an inverse-square model of gravity not only will yield elliptical orbits but is guaranteed to do so.

Newton went further still. He was able to encapsulate the previously mysterious workings of gravity within a simple, single formula that pertains not only to Earth and the Moon but to all bodies in the solar system and, indeed, to massive objects anywhere in the universe. This equation states that the strength of the gravitational attraction, or force F, between two massive bodies, m_1 and m_2, is proportional to the product of their masses and inversely proportional to the square of the distance (r) between them—or, more precisely, the distance between their centers of mass:

$$F = \frac{Gm_1 m_2}{r^2},$$

where G is the so-called gravitational constant.

It is perhaps worthy of mention that Newton had, by all accounts, a cantankerous personality.[4] At any rate, he engaged in a long and bitter priority struggle with Robert Hooke, who stated in his 1665 book *Micrographia* that gravity varies inversely with the square of distance.[5] But Hooke's argument, unlike Newton's, was never incorporated into a full-fledged theory of gravity. Hooke, moreover, did not know calculus, so he could never have attained the insight or range of results that Newton did.

Of course, Newton was by no means the first person to theorize about gravity. The phenomenon had, since prehistoric times, been apparent to people at least in a vague sense, and it had long served as a topic of rumination and debate among scholars. In the fourth century BCE, for example, Aristotle posited that objects tend to fall toward the center of the Earth, which he regarded as the center of the universe, at a speed proportional to their weight. Almost 2,000 years later, Galileo carried out experiments that contradicted one of Aristotle's key assertions: Under the influence of gravity, Galileo concluded, all objects fall at the same speed (neglecting the influence of air resistance or friction) and experience the same acceleration.

Newton took the next step—or what was actually a giant leap—several decades later, bringing in the mathematical scaffolding upon which to build our conception of gravity.

The period of time he spent at the family dwelling during the plague, centered around 1666, has since been called his *annus mirabilis*, or year of wonders. In that period, Newton laid down the foundations of calculus; made important advances in the field of optics, demonstrating the composite nature of light; and, of course, made his breakthrough on universal gravitation. "All this was in the two plague years of 1665 and 1666," he later wrote. "For in those days I was in the prime of my age for invention and minded Mathematicks and Philosophy more than at any time since."[6]

He continued to refine this research and combine it with his work on the laws of motion and other subject matters, eventually publishing the findings two decades later in his magnum opus, *Philosophiae Naturalis Principia Mathematica*, first published in 1687. In our own times, Stephen Hawking would call it "probably the most important single work ever published in the physical sciences."[7] In this masterwork, Newton introduced his three laws of motion, which lie at the heart of classical mechanics, as well as the derivation of his law of universal gravitation. (He did not cast his arguments in terms of calculus, which would have been the more concise and elegant way of doing so, because he wanted to present the discussion in terms that others could readily understand.) Newton showed in his *Principia* that Kepler's laws of planetary motion, which are based on observations of the solar system, follow mathematically from his (Newton's) own laws of motion and gravitation. He also provided, for the first time ever, a firm mathematical basis for understanding gravity, as well as a quantitative way of gauging its strength.

Newton's ideas have held up remarkably well indeed. In fact, all the navigational calculations in NASA's Apollo program were based on Newton's theory of gravity. The Apollo 8 mission, which flew in December 1968, brought humans to the Moon for the first time in order to orbit our nearest neighbor in space (though without any intention of landing there on this particular foray). "Astronauts were sent out in [a] spacecraft which circled the moon, orbited several times...and then came back to the Earth with essentially all the fuel gone, just relying on the validity of Newton's laws," the physicist Steven Weinberg commented.[8] During the return trip, while communicating with Mission Control in Houston, astronaut Bill Anders remarked, "I think Isaac Newton is doing most of the driving right now."[9] And, thanks to some skillful "driving," Anders and his companions—Frank Borman and Jim Lovell—landed safely on planet Earth, splashing down into the Pacific Ocean less than two days later.

Nevertheless, shortcomings of Newton's theory were uncovered, even though that theory still remains eminently useful. One problem was essentially philosophical: Although Newton could accurately calculate the effects of gravity and make accurate predictions, he could not offer any explanation for the mechanism behind it. In other words, he could not say how gravity worked—and this was a deficiency he openly acknowledged. "You sometimes speak of gravity as essential and inherent to matter," Newton wrote in a 1692 letter to Richard Bentley, a theologian and philosopher at Trinity College. "Pray do not ascribe that notion to me, for the cause of gravity is what I do not pretend to know."[10] Newton expressed similar sentiments in the "General Scholium," an essay he wrote that appeared at the end of *Principia*'s second edition: "I have not as yet been able to deduce from phenomena the reason for these properties of gravity, and I do not feign hypotheses."[11] Nor did Newton's gravity offer any explanation for Galileo's finding that all objects fall at exactly the same rate. Another puzzling feature of Newton's law, which did not sit well with some of his contemporaries, is that the force of gravity must somehow be transmitted instantly. Given that the force is proportional to the distance between two objects, if one object moves, the force exerted on the other is immediately and automatically adjusted, as if by magic—without any known or postulated means of conveying that change.

Even though Newton was aware of some of his gravitational theory's limitations, and knew that it left some important questions unanswered, he also recognized that it worked quite well. Rather than being drawn into philosophical debates over the fundamental essence of gravity, he was inclined to take a more utilitarian view: "It is enough that gravity exists and suffices to explain the phenomena of the heavens," he asserted.[12]

Such philosophical doubts about Newtonian gravity were successfully held at bay for nearly two centuries. However, a technical issue arose in the mid-1800s that could not be so readily brushed aside. Newton's laws of motion and gravitation could predict the motions of the planets in the solar system almost perfectly with one notable exception, that being Mercury. Its orbital motions departed slightly from the behavior that Newton's laws would have predicted. In 1859, the astronomer Urbain Le Verrier discovered what was going on with regard to Mercury's orbit around the sun: Mercury's perihelion—the point in the orbit where Mercury makes its closest approach to the sun—was not staying in the same place. With each revolution of the sun, the perihelion shifted a tiny bit, moving in the same direction Mercury followed as it revolved around the sun. This change in the location of the perihelion, which is, in turn, a change in the orientation of Mercury's orbit, is called a precession.

Every planet in the solar system experiences a perihelion shift, but only Mercury's does not comport with Newton's theory. The reason for Mercury's outcast status, it was later understood, owes to the fact that Mercury moves much more swiftly than the other planets and, being the closest to the sun, it is also subjected to the strongest gravitational effect. But Le Verrier calculated, back in 1859, that Mercury's perihelion precession was faster by 35 arc-seconds per century (with 1 arc-second being 1/3,600 of a degree) than would be expected based on Newtonian theory. In 1882, the mathematician Simon Newcomb refined that calculation, concluding that the extra precession—"extra" in the sense that it exceeded what Newton's laws could account for—was actually 43 arc-seconds per century.[13]

As to what was behind the disparity, astronomers suggested that Mercury's mysterious orbital behavior could be explained by an unknown planet that lay closer to the sun. This was, in fact, postulated by Le Verrier, who dubbed the hypothetical planet Vulcan. He suggested that, in lieu of Vulcan, the presence of a small group of undiscovered inner planets could also lead to the anomalous precession. But subsequent observations showed that no such planet, or group of planets, existed.

There was another possibility that was, in some sense, even more radical. Maybe Newton's gravity, which had ably served the world for so long, was wrong. Or at least not entirely correct. It was clearly good enough to explain most of the phenomena in the solar system, and indeed much of the universe, but there were certain phenomena—involving situations in which bodies were moving very fast or gravity is extremely strong—for which it could not account.

Perhaps a new theory of gravitation was needed—one that could reproduce the Newtonian version in situations where it had been proven to work well but could also handle the exceptional—and more extreme—instances where those same laws faltered.

That is where the problem lay until 1905, the year in which Albert Einstein made a very strong entrance onto the world stage. Up to that time, he was not especially well known. He was, in fact, toiling in obscurity as a patent clerk in Bern, Switzerland. In 1904, he had applied for a promotion from patent clerk third class to patent clerk second class. But his request was turned down by his supervisor, Friedrich Haller, who said that even though the applicant had "displayed some good achievements," his advancement would have to wait "until he has become fully familiar with mechanical engineering."[14]

In 1905, Einstein had a creative burst—his own *annus mirabilis*—that may never have been rivaled in the scientific world, except perhaps by Newton's outpouring in 1666. In that year, Einstein published four papers in the scientific journal *Annalen der Physik* (*Annals of Physics*), each of which fundamentally changed our understanding of the universe. The first paper, published in June 1905, was a milestone in quantum physics. In it, Einstein introduced the notion of the photoelectric effect, arguing that light does not just assume the form of smooth, oscillating waves but can also behave like discrete particles or bundles (*quanta*) of energy called photons.[15] (This paper was cited in the Nobel Prize in Physics that Einstein received in 1921.)

In a paper published in July 1905, Einstein explained the phenomenon of Brownian motion: Particles suspended in a liquid are constantly moving, he maintained, because of their continuous bombardment by unseen atoms. When viewed through this lens, Brownian motion offered clear evidence for the existence, and reality, of atoms, which people were unable to directly view at that time (and have only recently been able to observe through advanced microscopy techniques).[16]

In a September 1905 paper, Einstein presented the special theory of relativity,[17] and two months later, he revealed an extraordinary consequence of that theory.[18] Energy and mass are equivalent, he proclaimed, and their relationship is described in the following equation, arguably the most famous ever inscribed: $E = mc^2$.

Einstein was twenty-six when his first paper on special relativity came out, but he'd been thinking about some of the central issues for roughly a decade. What sent him on the path toward special relativity, he recounted in his *Autobiographical Notes*, were his reflections on a paradox he'd been contemplating since the age of sixteen: If he were to travel at the speed of light alongside a light beam, perhaps while riding in a speeding train, he would see a frozen light beam—a bizarre situation for which he could conjure up no plausible physical justification. "I should observe such a beam of light as an electromagnetic field at rest though spatially oscillating," he wrote. "There seems to be no such thing, however, neither on the basis of experience nor according to Maxwell's equations.... One sees that in this paradox the germ of the special theory of relativity is already contained."[19]

Two key principles lie at the heart of special relativity. The first, as Einstein explained, is that "every universal law of nature which is valid in relation to a coordinate system *C* must also be valid...in relation to a coordinate system *C'*, which is in uniform translatory motion relative to *C*."[20] As a consequence of this principle, if a person were sitting on a quiet, vibration-free train with the windows and shades completely closed, there is no experiment that he or she could perform to find out whether the train is moving at a constant velocity or at rest with respect to a nearby station.

The second key principle, according to Einstein, "asserts that light in vacuo always has a definite velocity of propagation (independent of the state of motion of the observer or the source of the light)."[21] Einstein was thus proclaiming that—for all uniformly moving observers and in all coordinate frames that are in constant relative motion—both the laws of physics and the speed of light must be the same.

Einstein further noted "that to speak of the simultaneity of two events had no meaning except in relation to a given coordinate system, and that the shape of measuring devices and the speed at which clocks move depend on their state of motion with respect to the coordinate system."[22] In that latter statement, Einstein is referring to two new phenomena that had been introduced to physics not long before special relativity's unveiling: time dilation (namely, that time passes more slowly for a clock that is moving within a particular frame of reference than for a clock that is standing still) and length contraction (a moving ruler will appear shorter than a stationary ruler when its length is measured along the direction of motion). Special relativity accounts for both of these notions.

Here, Einstein and his peers were challenging basic tenets that had been advanced by Newton more than two centuries before. In *Principia*, Newton asserted that space is a fixed, immutable backdrop in which the laws of physics invariably played out. Newton also advanced the notion of *absolute* time—a quantity that would yield the same measure for all properly functioning clocks, regardless of the speed or direction in which those clocks were moving. For Newton, the unwavering nature of space and time was more than just an interesting fact; it provided the foundation for his entire system of physics.

All that changed with special relativity. Space could no longer be considered unchangeable, given that spatial (or distance or length) measurements depended on motion. Time, as measured by a clock, was another quantity that varied with motion. Space and time, in other words, became *relative* concepts, with their values depending on how the observer who measures them happens to be moving.

Insights from relativity also cast doubt on a key feature of Newtonian gravity: that changes to its effects were felt immediately. But such alterations in gravity, special relativity tells us, could not be transmitted instantaneously or imparted at any speed faster than that of light itself. Indeed, special relativity tossed out the entire notion of absolute simultaneity, indicating that Newton's theory would, at a minimum, need to be modified, and perhaps eventually supplanted, in order to bring it into compliance with newly recognized laws of nature.

These ideas, to be sure, did not spring entirely from Einstein's imagination but actually had a long precedent. In a book published in 1632, Galileo introduced his own principle of relativity, describing a set of measurements involving flies, butterflies, and fish in a bowl that could be made onboard a ship when the ship is standing still. After doing that, he said, "have the ship proceed with any speed you like, so long as the motion is uniform and not fluctuating this way and that. You will discover not the least change in all the effects named, nor could you tell whether the ship was moving or standing still."[23] In this way, Galileo argued that uniform motion does not change the outcome of experiments.

There were, of course, more contemporary influences. Experiments in the 1880s by Albert Michelson and Edward Morley demonstrated that light always traveled at the same velocity and that the speed of light was not affected by the speed of the source. The physicist Hendrik Lorentz discovered how lengths contract. The mathematician Henri Poincaré also contributed many of the key concepts of special relativity. But Einstein offered a somewhat different and broader interpretation than all of them: He recognized that relativity theory applied to all physics, not just to electromagnetism or mechanics.

That said, Einstein acknowledged that the special theory of relativity was not the end of the story due to a central (and defining) limitation: Its focus was restricted to a "special" case—that of uniform or constant velocity—and was not broad enough to accommodate more arbitrary motions, specifically accelerated motions. In that way, the scope of the theory was artificially confined to certain phenomena in the natural and physical world, while being unable to have anything to say about other kinds of dynamical phenomena.

He came to realize in 1907, while writing a monograph for the journal *Jahrbuch der Radioacktivität und Elektronik* (*Yearbook of Radioactivity and Electronics*), "that all natural laws except the law of gravity could be discussed within the framework of the special theory of relativity. I wanted to find out the reason for this."[24] He was also determined to find a way of taking the principle of relativity that he had established in the idealized case (in which acceleration was not part of the picture) and generalizing it to systems that are not in uniform motion with respect to each other. That, needless to say, turned out to be a formidable undertaking.

A revelation, Einstein claimed, came later that year, while he was sitting in a chair in the Bern patent office. He experienced what is often called a eureka moment—being overcome, quite suddenly, by an idea he later described as "the happiest thought of my life." What he contemplated was this: "If a man falls freely, he would not feel his own weight." Continuing that line of reasoning, he said: "A falling man is accelerated. Then what he feels and judges is happening in the accelerated frame of reference. I decided to extend the theory of relativity to the reference frame with acceleration. I felt that in doing so I could solve the problem of gravity at the same time."[25]

Einstein's breakthrough is called the equivalence principle, and it is so named because it establishes an equivalence between acceleration and gravitation. A man falling, say from the roof of a house, would "not feel his own weight," as Einstein put it, nor would there exist for him, at least in his immediate surroundings, any gravitational field. The reason being that the falling man is accelerating and that acceleration would exactly counteract the sensation of weight he'd otherwise feel due to gravity.

Another way to think of it is to imagine this same person standing in a closed elevator. If he feels himself pulled toward the floor, he would have no way of telling whether the elevator was stationary and he was simply feeling the tug of gravity or whether the elevator was rapidly accelerating upward in a gravity-free environment (such as outer space). Similarly, if he released a stone from his hand and it ended up on the floor, he could not tell whether the stone fell under the influence of gravity or whether the stone had remained stationary and was hit by an upwardly accelerating floor. Once again, no experiment could be performed to distinguish between these two possible interpretations.

To carry this discussion a bit further, the downward force felt by the man in a stationary elevator is due strictly to his *gravitational* mass, which reflects the strength of gravity acting on him. The downward force he feels in the upwardly accelerating elevator is due to his *inertial* mass—which reflects the resistance of a body to being moved or, conversely, how fast a body accelerates when it is subjected to a given force. Einstein showed, in another way of stating the equivalence principle, that the gravitational mass always equals the inertial mass.

Again, Einstein was neither the only nor the first person to have thoughts along these lines. In the late 1500s, for example, Galileo carried out experiments that involved either dropping balls from the Leaning Tower of Pisa or rolling balls down an inclined plane, but the conclusion he drew was the same either way: The balls reached the ground at the same time, regardless of whether they had different weights or were composed of different materials.

Einstein, however, needed to reframe the problem in terms that resonated with him. And one thing he perceived upon recognizing the equivalence between acceleration and gravity is that if he could successfully extend his special theory to include accelerated motions, the general theory that he crafted would, in fact, be a theory of gravity.

And that's what he set out to do after his happy thought of 1907. But the answer did not come quickly or easily, Einstein acknowledged: "It took me eight more years until I finally obtained the complete solution."[26] One big hurdle he had to overcome during that interval was to master many new techniques from mathematics that he never imagined would be of use to him. The physicist Ivan T. Todorov summed up the situation this way: "By Christmas 1907, Einstein had all the physical consequences of the future gravity theory in his hands, yet he had another eight years to go and to appeal for mathematicians' help before arriving at the proper mathematical formulation of general relativity."[27]

Some unsolicited assistance came in 1908 from an unexpected quarter: Hermann Minkowski, who had been one of Einstein's mathematics professors at the Eidgenössische Technische Hochschule (ETH), a public research university in Zurich. In his college years, Einstein frequently missed classes, including many of Minkowski's lectures, which prompted the professor to describe his student as a "lazy dog [who] never bothered about mathematics at all."[28] However, as history has amply shown, Einstein was anything but lazy; he just had his own priorities. It seems likely, moreover, that few observers today would wish that the young Einstein had reordered his priorities and made class attendance, assiduous note-taking, and the punctual completion of homework assignments his chief goals in life.

Regardless of Minkowski's initial views of his former student, his thinking was clearly stimulated by Einstein's 1905 paper on special relativity. In a September 1908 lecture given at the eightieth annual meeting of the Society of Natural Scientists in Köln, Germany, Minkowski made this bold pronouncement: "From now onwards, space by itself and time by itself will recede completely to become mere shadows, and only a type of union of the two will stand independently on its own." He elaborated on this point in a separate lecture that he delivered to the Göttingen Mathematical Society: "What is being dealt with here is ... that the world in space and time in a certain sense is a four-dimensional non-Euclidean manifold."[29]

By introducing the notion of a four-dimensional spacetime, Minkowski offered a complete description of special relativity through geometry, while also laying the groundwork for the broader theory to come—general relativity. Space and time assumed equal footing in this new framework. A point in spacetime can be uniquely identified by specifying its four coordinates (x, y, z, t), which tell you when and where something, and indeed anything, is happening. That's similar to arranging to meet a friend in a building at the corner of 1st Street and 2nd Avenue on the third floor at 4:00 p.m., confident that you have provided the information needed for a successful rendezvous (although having no bearing on how that hypothetical encounter might go).

Relativistic effects that had previously seemed mysterious were easier to understand when viewed, and explained, through the lens that Minkowski provided, as they were essentially geometrical consequences of the unification of space and time that he had brought about. A point in spacetime, according to his terminology, was called an event, and the distance between two points (or events) was called a spacetime interval. If one person happens to be moving with respect to another, they will each come up with different values for distance and time as measured separately—owing to some of the phenomena mentioned before like length contraction and time dilation. But one thing the two observers will agree on is the interval or distance in four-dimensional spacetime.

That interval is a fundamental quantity whose value does not change, regardless of the frame of reference in which it is measured. Although the time component and space component may vary from frame to frame, the total length in spacetime stays the same. (As an analogy, consider a vector: a quantity, often denoted by an arrow, that is fully described by its magnitude and direction, centered at the origin of a two-dimensional x-y plane. As you rotate the vector, its x and y coordinates continually change, but its length does not.)

Minkowski provided a simple formula for determining the distance between two points in four-dimensional spacetime, which turns the Pythagorean theorem on its head: Motion in the time direction is *negative*. As the physicist Anthony Zee put it, "Imagine telling

Pythagoras that time has something to do with flipping a sign in his magical formula. You would have been certified as a total nut."[30]

In any case, here's the formula in Minkowski's four-dimensional spacetime for the distance (s) between two points, or actually the square of the distance:

$$s^2 = (\Delta x)^2 + (\Delta y)^2 + (\Delta z)^2 - (c\Delta t)^2,$$

where Δ (pronounced delta) represents the change in the x, y, z, and t coordinates in moving from one point to another. The term ct (the speed of light times time) actually represents a length. Moreover, if one chooses units that are "normalized"—such that the speed of light, c, is equal to one—the formula simplifies to

$$s^2 = (\Delta x)^2 + (\Delta y)^2 + (\Delta z)^2 - (\Delta t)^2.$$

The fact that a minus sign is placed in front of the time term is one reason the geometry of Minkowski spacetime is non-Euclidean. And, in contrast to what happens in Euclidean space, the hypotenuse of a right triangle can be shorter than one of its legs.

Let's consider a simplified graph in which a new variable X encompasses all three space directions, as represented on the horizontal axis, and the vertical axis is time. Distance in space equals X, whereas the distance in spacetime is $s = \sqrt{|X^2 - t^2|}$, where the straight brackets indicate absolute value, forcing the number to be positive regardless of whether X^2 or t^2 is greater.

We also know that X, the distance traveled in space, is the velocity multiplied by time. For a beam of light, that distance (X) is ct, but because we've set c equal to 1, $X = t$. Therefore, a beam of light, starting at the origin, would appear to travel along a 45-degree angle, passing through the point (1, 1) for instance. However, as the abbreviated formula above implies, the distance traveled by light in spacetime, $\sqrt{|X^2 - t^2|}$, is always zero, given that $X = t$. And that fact alone demonstrates that distance in Minkowski space is different from distance in Euclidean space.

Let's consider a point D in spacetime (X, 1) where X is greater than zero but less than one. We can see, using the simplified distance formula, that the length of the line segment AD is $\sqrt{|X^2 - t^2|}$. For a positive X whose value is less than one, that is always going to be shorter than the line segment AB, which has a length equal to 1. In contrast to the tenets of standard Euclidean geometry, the hypotenuse of this right triangle, AD, is *shorter* than AB, one leg of that same triangle.

Similarly, we can see that the length of CD, again equal to $\sqrt{|X^2 - 1^2|}$, is shorter than BC, which again equals 1. Logically then, the length of the sum of AD and CD, or $2\sqrt{|X^2 - 1^2|}$, is shorter than the straight-line distance of AC, which equals 2. So we've shown that the shortest distance between two points in Minkowski spacetime is not necessarily found along a straight line.

Once we can grasp that, much about special relativity falls into place. We can use this, for example, to understand the so-called Twin Paradox, a commonly discussed example in special relativity. As it's often presented, the tale begins with twin brothers on Earth. One stays on Earth; the other travels in a rocket ship at a high speed

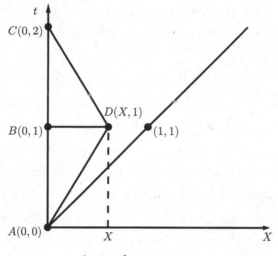

A triangle in spacetime

(though necessarily below the speed of light). He then returns to Earth only to find that his brother has aged significantly in the interim, while he has aged barely at all. That prediction is often referred to as a paradox because it seems a bit puzzling, but it is not really a paradox at all. Our figure shows why. The Earth-bound brother does not move and just travels along the time direction from A to C. The astronaut twin takes a high-speed rocket from A to D and then instantly hops onto another high-speed rocket that takes him from D to C and back to Earth. The rocketeer ages less simply because his journey through spacetime is much shorter. The faster his rocket travels, the greater the disparity will be. (However, upon a closer examination, it becomes clear that for the venturesome brother to return to Earth, he will have to accelerate—or, more specifically, decelerate—rather than continuing to travel at a constant velocity.

That lies outside the purview of special relativity. The general theory of relativity, which encompasses accelerated motions, is thus needed to gain a full understanding of this admittedly fanciful situation.)

The two twins start at the same place and end at the same place but take very different paths through spacetime, and the clocks they carry with them on these journeys reflect that disparity. The fact that their clocks measure different times, which is in turn reflected in their different ages, is no more paradoxical than saying that if two brothers drove from Los Angeles to San Francisco—one via the winding Pacific Coast Highway, the other by the much straighter Interstate 5— their respective odometers would measure different distances, even if they had the same starting and ending points.[31]

Returning to our original, not-so-paradoxical paradox, the fact that the twins age differently is no great mystery in light of Minkowski's insight. It is merely a consequence of geometry. And as seen through this example, geometry can also explain the phenomenon of time dilation: If you travel, time slows down, and the faster you travel, the slower it gets.

In addition, Minkowski also introduced the spacetime diagram— a geometric way of visualizing the structure of spacetime. Recall from our earlier discussion the simplified distance formula, $s^2 = X^2 - t^2$, in which X represents all three space coordinates (x, y, and z) and t represents the time direction (normalized so that the speed of light, c, equals 1). When $X = t$, you'll get a line at a 45-degree angle passing through the point (1, 1), and as the formula indicates, the distance from the origin to any point on this line is zero. This is the path that light always takes in Minkowski spacetime.

Similarly, one could draw another line at a 45-degree angle, this time passing through the point (–1, 1). Light starting at the origin and moving forward in time would travel along the V formed by these two 45-degree lines. And if we were to add a second space dimension (an x and y axis, with the t axis orthogonal in the vertical direction) and then rotate that V around the origin, we'd end up with a surface called a light cone.

In Minkowski spacetime, all light rays travel along the surface of the light cone—moving, of course, at the speed of light. These trajectories are called worldlines, a term coined by Minkowski that refers to the paths that any objects (including photons) trace in four-dimensional spacetime. Even if an object (such as a particle) is not moving in any space direction, it will still trace out a path in

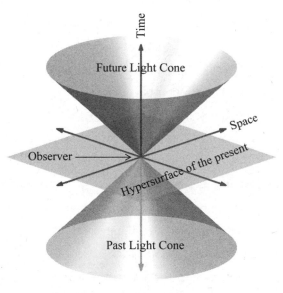

A light cone

spacetime, moving forward in the time direction. Any object following a path, or worldline, outside the light cone would necessarily
travel faster than the speed of light. That is forbidden in special (and
general) relativity, but is allowable in Newtonian mechanics where
there are no built-in speed limits.

If—contrary to Minkowski's formulation—you were to change
the negative sign in the distance formula to a positive sign, the light
cone would not exist anymore. A particle could then move with any
speed imaginable—though in our universe, that's simply not possible. No particle can move faster than the speed of light, and only
massless particles can move at the speed of light.

In other words, the minus sign that Minkowski curiously inserted
into the distance formula is, in fact, "related to the absoluteness of
the speed of light in relativity," explained the mathematician Mu-Tao
Wang. "In Newtonian gravity, space and time are absolute and independent." Whereas in relativity, space and time are intertwined,
Wang added, "and only the speed of light is absolute."[32]

Minkowski's ideas, like other big conceptual leaps, did not win immediate acceptance, despite the fact that by recasting special relativity into a four-dimensional theory, he was able to explain phenomena
that could not be so readily grasped before. The great mathematician
Poincaré, for instance, did not see the value of Minkowski's reformu-
lation. Although "it would be possible to translate our physics into
the language of geometry of four dimensions," Poincaré asserted in
1908, "to attempt this translation would be to take great pains for
little profit." Minkowski, however, took a different view, insisting

that four-dimensional spacetime is essential for a full understanding of physical laws.[33]

Einstein—who is often wrongly given credit for inventing the notion of four-dimensional spacetime—was not initially impressed with Minkowski's contribution either. He dismissed it as "superfluous learnedness"[34] and complained to a friend that ever "since the mathematicians have invaded the relativity theory, I do not understand it myself anymore."[35]

Of course, many scientists view things differently now. The mathematician (and mathematical physicist) Roger Penrose has said that special relativity was not a complete theory until Minkowski demonstrated that the theory was best understood within the context of four-dimensional spacetime.[36]

But at the time, Einstein and other physicists were leery of pronouncements like the one Minkowski had advanced in 1908. "Einstein in the beginning of his career distrusted mathematics and considered the mathematical formulation of a physical event as the mere form in which a phenomenon is described, which does not touch on its substance," explained the mathematician and physicist Cornelius Lanczos. "He employed mathematics only so far as was necessary to bring out the essence of the physical idea."[37]

Einstein, unsurprisingly, was able to create his theory of special relativity in 1905 while treating time and space (of the three-dimensional variety) separately, without the benefit of the ideas that Minkowski put forth three years later. And for several years after that, proceeding in the spirit Lanczos described, Einstein continued to pursue three-dimensional versions of gravity. But those efforts eventually hit a dead end. By 1912, he came to see that he would need to adopt a four-dimensional framework that incorporated time—as well as some of the mathematical methods that Minkowski had utilized—in order to move forward in his development of a gravitational theory.

That realization may seem obvious now when you consider that in a dynamic universe, in which matter is moving and its distribution is constantly changing in time, an adequate description of gravitational fields would necessarily require three dimensions of space and one of time. But up until that moment, it was not obvious to Einstein or to many, if any, of his peers. Yet one could say that Einstein's (belated) recognition that four-dimensional spacetime was the appropriate and indeed necessary setting for general relativity was as important to the establishment of that theory as was his embrace of the equivalence principle—his "happiest thought" of five years earlier. This thought—which embraced the notion that space and time were fully coupled to each other (and, as we shall soon see, to gravity as well)—might have been even happier.

Einstein offered a belated tribute to his former teacher in the introduction of a March 1916 paper, "The Foundation of General Relativity," in which he wrote: "The generalization of the theory of relativity has been facilitated considerably by Minkowski, a mathematician who was the first one to recognize the formal equivalence of space coordinates and the time coordinate, and utilized this in the construction of the theory."[38]

Einstein showed just how far his thinking had changed regarding Minkowski's work in a popular book he wrote in 1916. "The non-mathematician is seized by a mysterious shuddering when he hears of 'four-dimensional' things, by a feeling not unlike that awakened by thoughts of the occult. And yet there is no more commonplace statement that the world in which we live is a four-dimensional spacetime continuum." He then went on to state that without the framework provided by Minkowski, "the general theory of relativity... would perhaps have gotten no farther than its long clothes"—the latter term referring to infant garments that are, in some translations, called diapers.[39]

Minkowski, sadly, did not live long enough to see that his work had been fully embraced. Nor did he have the chance to explore a possibility he had raised in 1908—to extend the theory of relativity to include gravitation.[40] He died of a ruptured appendix in January 1909, less than three months after the famous "Space and Time" lecture he delivered in Köln. Shortly before his death, Minkowski sketched out a way that such a gravitational theory might be constructed. Although he was not able to develop this idea further, he opened a pathway to it. It's hard to know how far Minkowski would have gotten if he had more time, noted the mathematics historian Leo Corry. "What we know for certain is that his choice of the four-dimensional formulation... turned out to be absolutely fundamental for the possibility of formulating any relativistic theory of gravitation."[41]

It wasn't until 1912—or possibly a bit earlier, as some historians contend—that Einstein recognized four-dimensional spacetime was the requisite setting for a gravitational theory, but that, alone, was not enough to get him to the finish line. He reached another major conclusion, likewise anticipated by Minkowski, in that same year—namely that he would have to move beyond Euclidean geometry to reach his goal.

This realization stemmed from another thought experiment, which Einstein carried out that year while he was a professor at the German University of Prague. Special relativity pertained to objects moving at constant velocities, but this time—in his attempt to move beyond the confines of that theory—Einstein considered an object in accelerated motion, not falling but instead spinning at a high but unvarying speed. A system of this sort—a wheel or so-called rigidly rotating disk—is undergoing uniform acceleration because the motion of every point on the wheel (save for the absolute center) is continually changing direction. Given that moving rulers shrink in the direction of motion, Einstein reasoned that the wheel's spokes (and hence the circle's radius, r) would not be affected by length contraction because their orientation is perpendicular to the wheel's motion. But the wheel's circumference, which lies along the direction of motion, would shrink. Therefore, a standard principle of Euclidean geometry—the fact that a circle's circumference equals $2\pi r$—would no longer hold. In other words, the geometry of this system, which is undergoing acceleration, would be non-Euclidean. Whereas Euclidean geometry—which describes flat space, where parallel lines never cross—is sometimes referred to as plane geometry, non-Euclidean geometry pertains to curved space or curved spacetime, where parallel lines can and do converge, just like the meridian lines on a globe that meet at the north and south poles. Einstein knew from the equivalence principle that if accelerated motion can lead to a curved geometry, then gravity would necessarily do the same. Putting it all together, he reached a logical, although still remarkable, conclusion: The geometry of spacetime in the presence of a gravitational field is not Euclidean. And taking it a step further, he realized that gravity is not a force in the way that Newton pictured it; gravity is nothing more, or less, than a consequence of the curvature—or geometry—of spacetime.

What that means is that planets in our solar system travel around the sun in elliptical orbits not because of the sun's gravitational pull, but because they are traveling along a surface, or spacetime, that has been deformed by the more massive object, the sun. The path taken by these planets—and, indeed, by every object or particle acting solely under the influence of gravity—is called a geodesic. The geodesic looks like a straight line in flat Minkowski space, but the trajectories prescribed by the gravitational laws Einstein was formulating can appear entirely different in a curved spacetime. It may, for instance, represent the shortest distance in the space direction but not in the time direction. "It's now up to every particle in the universe to follow the best path in this curved environment," Anthony Zee said. "This explains why gravity acts indiscriminately on every particle in exactly the same way."[42]

With that revelation, Einstein had hit upon a conceptual scheme with which to geometrize gravity, just as Minkowski had geometrized special relativity. However, Einstein's great realization was not the end of the story but merely an inflection point. He now needed to shift gears and come up with a mathematical formulation that spelled out the precise connection between spacetime curvature and the associated gravitational effect. Therein lay the rub. The path ahead,

Einstein explained, "was thornier than one might suppose because it demanded the abandonment of Euclidean geometry,"[43] and for him that meant abandoning the mathematics he knew and instead taking a deep dive into the strange, unfamiliar terrain of curved spacetime.

Lacking a background in non-Euclidean geometry, Einstein found it difficult to make much headway on his agenda. Yet he realized, nevertheless, that he would need new theoretical (and mathematical) tools in order to devise a theory, as he wrote in a July 1912 letter, "in which the equivalence of inertial and gravitational mass finds expression"[44]—a crucial condition to satisfy in going from the domain of special to general relativity.

Fortunately, he had a friend to turn to—Marcel Grossmann, a former classmate at the ETH who had since become an accomplished geometer. Grossmann had helped Einstein in college with some mathematics coursework, when Einstein's mind was on other subjects. Einstein desperately wanted his friend's help again—not with homework this time, but with something bigger and much more ambitious.

"You must help me," he implored, "or else I will go crazy."[45]

= 2 =

Finding a General Path
Forward

"This could be the plot of a novel," wrote Jürgen Jost, director of the Max Planck Institute for Mathematics in the Sciences based in Leipzig. "The main character is a shy, sickly young mathematician living in poor conditions at a German university in the middle of the 19th century."[1] Before attending college, this youth had been encouraged by his father, a Lutheran minister, to study theology at the University of Göttingen. But after attending some mathematics lectures, he secured his father's permission to switch to the philosophy program so that he could pursue mathematics. Nevertheless, he continued to study the Bible and even tried, at one point, to prove the mathematical correctness of Genesis, the first book of the Bible.[2]

After earning a PhD at Göttingen, our hero began working toward a habilitation degree, which was needed to obtain a professorship at a German university. Tradition held that such a candidate had to propose three topics for a habilitation colloquium. The mathematician had no difficulty choosing the first two topics, which were drawn from areas in which he'd already made some technical contributions. The third topic on his list was of a somewhat amorphous philosophical nature, but he didn't worry about it since the faculty almost always selected one of the first two choices. In this case, however, he was asked to give a talk on the last option—the one about which he was, by far, the least prepared to speak. Nevertheless, his presentation didn't just turn out well. It ended up changing history.

The young mathematician in question, Bernhard Riemann, proved himself to be more than up to the challenge thrust upon him. And on June 10, 1854, he gave a talk at his university with a rather grandiose title: "On the Hypotheses Which Lie at the Bases of Geometry." In his presentation, Riemann unveiled an entirely new way of thinking about the curvature of higher-dimensional space, thereby laying the foundations of modern geometry—a key portion of which, Riemannian geometry, has been vital to theoretical physics in addition to its lasting importance to mathematics. In the years since Riemann addressed the audience in Göttingen, noted Jost, "subsequent generations of mathematicians [have] worked out the ideas outlined in the brief lecture and confirmed their full validity and soundness and extraordinary range and potential."[3]

Riemann's ideas were so far ahead of his time that his academic advisor, the great mathematician Carl Friedrich Gauss, might have been the only one in attendance able to fully grasp, as well as appreciate, the contents. Gauss himself was a pioneer in the study of non-Euclidean surfaces, and he is considered one of the greatest mathematicians in history—having made major contributions to many of the areas of mathematics that are still being pursued today. Like others of his contemporaries in the early 1800s, he grappled with Euclid's fifth postulate, also called the parallel postulate. Two lines in the same plane are considered parallel if they always maintain the same distance apart and never cross. The fifth postulate basically maintains that if a line segment intersects two lines in such a way that the sum of the interior angles on one side of the segment is less than two right angles (i.e., less than 180 degrees), then the lines will eventually meet on that same side of the (intersecting) line segment. Another (simplified) way of saying that is that non-parallel lines in a plane eventually converge whereas parallel lines never do.

Many mathematicians had long felt the postulate should actually be provable using Euclid's first four postulates. No one had succeeded since the days of ancient Greece, and Gauss wondered whether it might not be provable after all. Perhaps, Gauss mused, it would be worth considering the possibility of two-dimensional spaces or surfaces in which the fifth postulate did *not* hold. His idea was that objects sitting within a particular space weren't the only things that could be curved; space itself could be curved. Understanding such spaces would require a more general form of geometry—one that was not limited to describing flat surfaces or the events taking place within them. Gauss expressed this "heretical" opinion in an 1824 letter to another German mathematician, Franz Taurinus, and Gauss subsequently began to explore non-Euclidean geometry at roughly the same time that these ideas were independently being developed by two other pioneers in this emerging branch of mathematics, János Bolyai and Nikolai Lobachevski.[4]

All of this may sound rather esoteric, but you may be quite familiar with some features of non-Euclidean space without even realizing it. On the surface of a sphere, longitudinal lines that may appear parallel do indeed converge (at the north and south poles), and the sum of the angles of a triangle add up to more than 180 degrees. By contrast, on a two-dimensional, saddle-shaped surface (such as a hyperboloid), seemingly parallel lines diverge, whereas the sum of the angles of a triangle add up to less than 180 degrees. This is in contrast, of course, to a flat (Euclidean) plane where parallel lines remain parallel, and the angles of a triangle add up to exactly 180 degrees.

 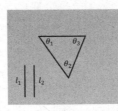 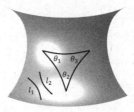

$\theta_1 + \theta_2 + \theta_3 > 180°$ $\qquad$ $\theta_1 + \theta_2 + \theta_3 = 180°$ $\qquad$ $\theta_1 + \theta_2 + \theta_3 < 180°$

Spherical (positive curvature) $\quad$ Euclidean (zero curvature) $\quad$ Hyperbolic (negative curvature)

Lines and triangles on surfaces with positive,
zero, and negative curvature

Of particular interest to Riemann was the theory of surfaces that Gauss advanced in an 1827 paper, known in English as "General Investigations of Curved Surfaces." In the previous century, Leonhard Euler had undertaken pioneering work on the curvature of surfaces, but the curvature properties established by Euler depended on the way a surface sits, or is embedded, in three-dimensional space—*embedding* here referring to the precise manner in which a mathematical structure is incorporated into a larger structure, such as a one-dimensional circle sitting in a two-dimensional plane. In this newer work, Gauss invented the notion of intrinsic geometry—the idea that an object or surface has its own curvature (which came to be known as Gaussian curvature) that is independent of how and where it may be sitting in space. A sphere has positive curvature, and a hyperboloid has negative curvature. A surface's intrinsic curvature, moreover, can be determined solely on the basis of measurements taken within the surface itself; no outside (or extrinsic) views are required.

Gauss's idea, as expressed in his Theorema Egregium (which translates to "Remarkable Theorem"), is that the curvature of a two-dimensional surface can be fully ascertained by measurements of the distance and angles between points on that surface. That curvature will not change, moreover, even if the surface is bent, so long as it is not stretched, compressed, or torn. Nor will the curvature be affected by how the surface is situated within two- or three-dimensional space.

Let's start with the example of a cylinder—of radius one, for the sake of simplicity—which we might picture as a can standing upright on a table with both its top and bottom portions removed. Imagine that there's a bug on the can, resting roughly halfway between the top and bottom. If, on the one hand, the bug were to travel directly up or down, it would follow a straight-line path of the smallest possible curvature (0). If, on the other hand, the bug chose to travel in the orthogonal direction—following a circular route around the can—that would be the path of greatest possible curvature. (The curvature of a circle equals $1/r$, r being the radius of curvature, which in this example would have a value of 1/1 or 1.) The Gaussian curvature at any point on a surface is the product of these two so-called principal curvatures, the smallest and the largest. In the case of an open-ended can or cylinder, the product of the principal curvatures at any point on the surface equals 1 × 0, which means the Gaussian curvature of a cylinder is 0.

Expressing this in equivalent terms, a cylinder is flat. Though you might think a cylinder looks kind of round, that idea actually makes sense because one can take a flat, rectangular piece of paper and roll it up into a cylinder. As Gauss saw things, even though a cylinder consists of a rolled-up piece of paper, its curvature is the same as that of a flat (unrolled-up) piece of paper—0, in other words. Again, that makes sense because the distance between any two points on the cylinder—as measured along the surface—remains unchanged, regardless of whether the paper lies flattened out on a table or rolled up into a tube.

Similarly, if our cylinder were made of a pliable material, like that of a garden hose, it could be curled up again—until its ends meet—to

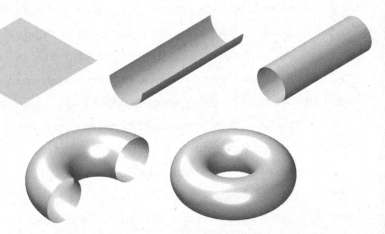

A flat piece of paper transforms into a cylinder and then into a donut

form a donut or torus. And the torus would have the same intrinsic curvature as the cylinder or a flat piece of paper: 0 in all cases.

A two-dimensional sphere, as another example, has a different intrinsic curvature. Let's imagine that our sphere again has a radius of 1 and looks like a globe. If we start from a point on the equator, traveling around the equator would take us on a path of maximal curvature, which is equal to 1. Traveling in the orthogonal direction, on a circle that passes through the north and south poles, is also a path of curvature 1. That same reasoning would, in fact, apply to any point on the sphere, meaning that its Gaussian curvature is everywhere 1 × 1 or 1.

The advent of Gaussian curvature can be considered a, if not *the*, starting point of modern geometry. Gauss's insights on the inner properties of curved, non-Euclidean spaces were both tremendous and profound, but they were limited to two-dimensional spaces. Extending that work—and describing the intrinsic curvature of spaces having three, four, and higher dimensions—turned out to be a challenging problem, one that was solved by Riemann in 1854 when he ushered in a new form of geometry that eventually bore his name.

Central to Riemann's reimagination of geometry was the concept of a manifold, which he introduced to describe a space of any dimension or indeed of arbitrary n dimensions. A more rigorous definition of a manifold, which comes close to how we view the concept today, was introduced almost sixty years later—in a 1913 book, *The Idea of a Riemann Surface*—by the mathematician Hermann Weyl. A manifold, according to the current picture, is a smooth space or surface that appears flat, or Euclidean, on a tiny (infinitesimal) scale, but can be curved when viewed on a larger scale.

The planet Earth, for example, appears flat on a very small scale but resembles a sphere as seen from outer space. A sphere is a special kind of manifold—one that has constant and positive curvature everywhere on its surface. A manifold needs to be smooth, with no tears nor sharp or jagged edges. Indeed, a manifold is guaranteed to be smooth, owing to the insistence that every tiny patch of it appears flat when taken individually. But a manifold need not have uniform curvature. Its curvature can vary from point to point, smoothly and gradually, and it can also change depending on the direction in which a bug, for instance, chooses to travel from a given point.

Each point on a four-dimensional manifold, for example, can be specified by four coordinates, and similarly n coordinates are needed to specify a point on an n-dimensional manifold. Riemann devised a construct, called the metric tensor, for determining the distance between any point on the manifold and another point nearby. In a flat Euclidean plane, distance can be measured simply, just using a ruler, or calculated quite easily using the Pythagorean theorem. But in higher-dimensional Riemannian space, in which the curvature on a manifold can vary from point to point, the usual Pythagorean theorem does not apply. Working out distances in curved space, consequently, becomes a much less straightforward proposition. A metric is thus needed to define the distance function and thereby facilitate such determinations.

From the distance measurements afforded by the metric, the curvature at every point on the manifold can be determined. For example, in four dimensions (the appropriate choice for describing spacetime in special and general relativity), the metric tensor is a four-by-four array, or matrix, with sixteen terms altogether, only ten of which are independent. Ten numbers are thus needed to describe the manifold's properties at each point, as well as to characterize the curvature of the manifold as a whole. All of this information is contained within the metric, which tells you the geometry of that space *locally*—that is, within a small area around a point. But from the distance calculations enabled by the metric, you can determine the geometry of that space in a *global* way. So the metric can, from a mathematical perspective, tell you practically everything there is to know about that space.

The value of the tensor's ten individual numbers associated with a particular point can change, depending on the chosen coordinate system, but the distance calculated between any two points—by means of a technique that relies heavily on the Pythagorean theorem—is independent of the coordinate choice. That accorded with Riemann's view that physical laws and meaningful properties of space should have nothing to do with the coordinates we choose or with the frame of reference that an observer happens to occupy. This inherent feature of Riemannian manifolds, called general covariance, incorporates the equivalence principle discussed in the previous chapter, and it will acquire heightened importance in our ensuing discussion of gravity.

The Riemann curvature tensor, which can be derived from the metric tensor, was introduced in Riemann's 1854 lecture, and he developed it further in a paper he submitted in 1861 for a prize to the Academy of Sciences in Paris (though he did not win).[5] Riemann had recognized that in manifolds of dimensionality greater than two, curvature is far too complicated to be captured by a single number. The Riemann curvature tensor is, instead, a rather elaborate array of numbers and functions (of four variables) that can, within a compact format, fully describe the curvature of multidimensional manifolds.

This tensor is classified as rank 4, denoted by four indices or subscripts—letters that correspond to the number of directions, and hence the dimensionality, of the array itself. The tensor's rank or dimensionality can be different from the dimensionality of the underlying space. A tensor of rank r in n-dimensional space would have n^r components. The Riemann curvature tensor in four-dimensional space, therefore, would have 256 (4^4) separate components, twenty of which are independent and thus contain twenty different pieces of information. A rank 3 tensor can be pictured as a three-dimensional array of numbers and functions that assume the shape of a cube. A matrix—a rectangular array consisting of rows and columns with two built-in directions—is a tensor of rank 2 and has two corresponding indices. A vector—which, by definition, has just a single direction assigned to it—is a tensor of rank 1. A number, or scalar, is a tensor of rank 0; having neither indices nor directionality, it just describes the magnitude of a particular quantity at a given point.

But the most important (and most ingenious) feature of the Riemann curvature tensor is that even though its components can change under a shift in coordinate systems, the tensor transforms in a well-behaved, linear fashion, which means that essential information can still be extracted. This property accorded with Riemann's view that physics should be independent of the choice of coordinates, and it was this same property that attracted Albert Einstein many decades later in his quest to build a gravitational theory around the equivalence principle.

With the invention of the curvature tensor, Riemann took much of the mystery out of multidimensional manifolds, which had previously been a kind of terra incognita. He also provided the mathematical framework that was applied many years later to understanding a very special four-dimensional manifold, one that represents the spacetime we inhabit—our universe, in other words.

It is perhaps stating the obvious to say that Riemann was a superb mathematician, unquestionably one of the great geniuses of the nineteenth century. In addition to his seminal work in geometry, he helped generalize the study of polynomial equations, while making major contributions to complex and real analysis, the theory of functions, and number theory. The Riemann hypothesis—a problem pertaining to the distribution of prime numbers that he posed in 1859—can surely be considered one of the (if not *the*) deepest open problems in all of mathematics. And mathematicians are currently grappling with many other concepts brought to the fore by this truly original thinker.

That said, Riemann didn't restrict himself to mathematics alone. He wondered, for instance, how the novel geometric principles he was developing might apply to space writ large, which is to say the universe itself. He alluded to such ideas in his 1854 lecture, contemplating the "physical reality upon which space is founded" and also questioning "the validity of the axioms of geometry in the infinitely small." In order to advance our understanding of the universe, he said, we should start with the foundations laid down by Newton, while making sure that our outlook is "not hampered by traditional prejudices," nor should we "be hindered by too narrow a view of the possibilities." At the time of his habilitation, however, those questions would have to wait: "This," he noted in that lecture, "takes us into the realm of another science, physics, which the nature of today's occasion does not allow us to explore."[6]

Decades ahead of his time, Riemann contemplated the possibility that the structure of space could be affected, and curved, by the presence of matter. He mused, moreover, that the space we inhabit may, in fact, be a curved manifold. He also had ambitions to devise a unified theory that stitched together the laws of electricity, magnetism, light, and gravity.[7] But he was unable to fulfill that dream. Riemann suffered from poor health throughout his life, and he died of tuberculosis in 1866 at the age of thirty-nine.

What would have happened if he'd lived at least a couple of decades longer and witnessed the results of the Michelson–Morley experiments, which affirmed the constancy of the speed of light? It's possible that he might have discovered special relativity, which would have fit naturally into the framework of Riemannian geometry, since Albert Michelson and Edward Morley had demonstrated that the speed of light was independent of both coordinate systems and the motion of a light source relative to an observer. Given Riemann's deep grounding in the principle of covariance, which is intimately tied to the equivalence principle, it's possible that he may have carried on from there, embarking on a search for a more general theory. Sadly, he never had the chance. Instead, those tasks were taken up—some four decades after Riemann's death—by another scientist, and it is to his story that we shall now turn.

In August 1912, Einstein arrived in Zurich to become the head of the theoretical physics department at the ETH. But he had another reason, apart from his new appointment, to return to his alma mater: His friend Marcel Grossmann was a professor of geometry there, and that happened to be a subject in which Einstein desperately needed help.

Einstein had come to the realization that Euclidean geometry could not be applied to the theory of gravity he was trying to build, in which frames of reference that are accelerating with respect to one another are equivalent.[8] That is, the general theory he was striving for, unlike special relativity, would not be restricted to objects moving in straight paths and at constant speeds, but would apply to all kinds of motions, including objects that undergo accelerated motions along curved paths. And the equations that describe such motions, moreover, would have to be the same for all observers—not only those that are in uniform motion with respect to one another.

Those were the contours of the problem he faced, yet he was not familiar with the non-Euclidean framework that would best serve his purposes in crafting a gravitational theory. In particular, Einstein sought a kind of geometry that had the property of general covariance—one in which the selection of coordinates would be immaterial, having no bearing on the physics. That made perfect sense, of course, since coordinates really just offer ways of labeling things. A parabola, for instance, could be centered around the origin of a particular graph. If the coordinates shifted, and the origin moved, say to the right or left, the parabola itself would not change in the slightest; all that would change was the way we labeled it. The physics of a situation, similarly, should not depend on the arbitrary choices we make about how to label things.

Such a property, he recognized, was essential for his new theory of relativity to earn the label *general*, meaning that the laws of gravitation it described would be identical for all observers, regardless of their coordinate choices or whether they happened to be surveying events from within accelerated frames of reference. The challenge was daunting, he recognized—arguably the hardest he'd ever taken on—as he called the attempt to express physical laws without coordinates "equivalent to describing thoughts without words."[9]

Einstein did not realize at the time that a form of geometry endowed with this very feature had been developed decades before. He asked for Grossmann's advice, and "the next day," according to the physicist and science historian Abraham Pais, "Grossmann returned…and said that there indeed was such a geometry, Riemannian geometry." He warned Einstein, however, that the differential equations of Riemannian geometry were nonlinear, which could result in "a terrible mess which physicists should not be involved with."[10] Einstein was not deterred by these words of caution, as he had already come to see that the gravitational field equations should, for physics reasons, be nonlinear—even if that meant they would be burdensome to manage.

Einstein quickly recognized, moreover, that Riemannian geometry was indeed "the correct mathematical tool," Pais said, and that "abrupt realization was to change his outlook on physics and physical theory for the rest of his life."[11] He also recognized that—in crafting a theory of gravitation—he would have to combine Riemann's ideas about curved space with Minkowski's concept of four-dimensional spacetime.

Although Einstein was able to take advantage of preexisting mathematical ideas, the application of Riemannian geometry, intertwined with the notion of spacetime, was not enough, in itself, to give Einstein the theory he sought. There was still considerable work to be done—more than three years' worth, as it turned out.

The development of Riemannian geometry had not started nor ended with the 1854 lecture, although it nearly did. Given that Riemann never published his lecture before his death, and that Gauss died in 1855, "Riemann nearly took these daring ideas with him to his early grave," wrote the math historian David E. Rowe.[12] Fortunately, the mathematician Richard Dedekind saw to it that Riemann's lecture was published in 1868. In a paper published one year later, the mathematician Elwin Bruno Christoffel built upon Riemann's work, developing the Riemann–Christoffel curvature tensor that subsequently became the standard mechanism for encapsulating curvature information.[13] The Italian mathematicians Gregorio Ricci-Curbastro and Tullio Levi-Civita carried these ideas even further, inventing tensor analysis or tensor calculus, as it came to be known—a methodology, which Ricci called absolute differential calculus, for manipulating tensors in a Riemannian manifold. These were the tools on which Grossmann briefed Einstein.

Ricci and Levi-Civita had generalized Riemann's ideas, translating them into a tensorial format that could be applied in a range of contexts. That format turned out to be an especially convenient one in which to write the equations of physics for curved spaces—a fact that

Minkowski became aware of well before Einstein did. Minkowski not only contributed the notion of four-dimensional spacetime and explained its utter importance, but he also recognized, and demonstrated, that spacetime was best described via the mathematics of tensors, several years before Einstein came around to that viewpoint.

Although Riemannian geometry was basically what Einstein needed to describe spacetime in general relativity, he still had to make one crucial modification, which relates to the metric. In Riemannian geometry, the metric is always of a type called positive-definite—meaning that the distance between any two points on a Riemannian manifold is always positive, unless those two points happen to be one and the same, in which case the distance between them is zero. But the distance in Minkowski spacetime is not always positive, and the Minkowski metric is thus classified as indefinite rather than positive-definite. Any point along the light cone, for instance, is considered zero distance from the origin, no matter how "far" out from the origin you go. And motion in the time direction—owing to the insertion of the minus sign in the distance formula—counts as negative, rather than positive, in terms of the distance traveled.

Some of this can perhaps be visualized by looking at the metric for Minkowski spacetime (in which the first diagonal component, corresponding to time, is negative and the three other diagonal components, corresponding to the x, y, and z spatial directions, are positive):

$$\begin{bmatrix} -1 & 0 & 0 & 0 \\ 0 & 1 & 0 & 0 \\ 0 & 0 & 1 & 0 \\ 0 & 0 & 0 & 1 \end{bmatrix}.$$

If the right coordinate system is chosen, the metric for four-dimensional Euclidean space can reduce to an array of 1's and 0's, making it identical to the metric for Minkowski space, except that travel in the time dimension is conferred positive (rather than negative) length:

$$\begin{bmatrix} 1 & 0 & 0 & 0 \\ 0 & 1 & 0 & 0 \\ 0 & 0 & 1 & 0 \\ 0 & 0 & 0 & 1 \end{bmatrix}.$$

These numbers describe a point with coordinates for time, t, and three space axes, x, y, and z. In the example described here, the coordinates for t, x, y, and z have been given the values (1, 1, 1, 1), and the metric tells us how far that point is from the origin (0, 0, 0, 0). If the metric you're looking for is symmetric, as it is in this example, it's possible to find a way of expressing things—a way of orienting all of the axes, you might say—so that all of the matrix elements are 0's except along the diagonals. (This, in linear algebra terms, comes down to finding the right *basis*.) For a metric of this special (symmetric) type, distance can be computed easily—just by adding up the diagonal elements.

In Euclidean space, the distance, s, from the origin to the point labeled as (1, 1, 1, 1) is the square root of the sum of the diagonal elements:

$$s^2 = t^2 + x^2 + y^2 + z^2 = 1 + 1 + 1 + 1 = 4$$
$$s = 2.$$

Similarly, in Minkowski spacetime, the distance from the origin to the point in question (−1, 1, 1, 1) can be determined by taking the square root of the sum of the diagonal elements:

$$s^2 = -t^2 + x^2 + y^2 + z^2 = -1 + 1 + 1 + 1 = 2$$
$$s = \sqrt{2}.$$

General relativity, as we've established, involves an extension of special relativity from flat (Minkowski) spacetime into curved spacetime, and the correct curved spacetime for this theory happens to be a four-dimensional Lorentzian manifold, named after relativity pioneer Hendrik Lorentz. A Lorentzian manifold represents a kind of blending of Minkowski and Riemann. Whereas a Riemannian manifold, locally (that is, on a very small scale) looks like Euclidean space, a Lorentzian manifold, locally, looks like Minkowski space—endowed with a different brand of flatness. A Lorentzian manifold, more specifically, is a generalization of a Riemannian manifold (classified as a pseudo-Riemannian manifold) because, as with Minkowski spacetime, the positive-definite condition is relaxed. The distance between any two points on a Lorentzian manifold does not have to be positive in all cases.

Accordingly, the metric for a Lorentzian manifold—when written in its simplest, symmetric form—again can have just four diagonal elements: One corresponds to the time direction with a negative sign in front, and the other three, which correspond to the space directions, are positive. However, unlike the metrics shown in the previous example, in which all the elements were either 1 or –1, the four diagonal elements of a Lorentzian metric don't have to be just numbers; they can instead be four *functions*—one function that yields a negative output and three others that yield positive outputs.

So how might the metric tensor tell us the distance between two points, *A* and *B*, lying near each other on the same curve? At every point along that curve, the metric gives you a number; in other words—as stated above—it defines a function on that curve. If, applying calculus, you integrate the function along the curve from A to B, you can also find out the length of the curve, which is exactly what a metric is supposed to do.

It is no coincidence that Minkowski spacetime (the natural setting of special relativity) is a special case of a four-dimensional Lorentzian manifold (the natural setting of general relativity). What it's telling us is that general relativity reduces to special relativity under special circumstances—namely when spacetime is flat and when gravity and acceleration do not enter the picture. This can be stated in other terms, which turns out to be a different, as well as a stronger, version of the equivalence principle introduced in the last chapter: In a small neighborhood around any point in spacetime, or along a geodesic, the laws of physics reduce to those of special relativity. This harkens back to our earlier definition of a manifold—namely that it has to be flat at each point on a local, or infinitesimal, scale. *Flat* in this case means that it looks like Minkowski spacetime in a tiny patch surrounding each point.

The upshot of all this is that Einstein's efforts to expand his theories from the confines of special relativity to the more complex milieu of general relativity were reflected by a parallel progression in geometry—moving, or generalizing, from flat Minkowski space to a curved Lorentzian manifold, which was itself a modified version of a Riemannian manifold.

After Einstein had selected a suitable geometry for curved spacetime, he turned to the mathematical methods of Ricci and Levi-Civita. Their techniques provided a way of differentiating non-flat space, while at the same time ensuring that the results of that differ-entiation do not depend on the choice of coordinates. The approach turned out to be tailor-made for formulating the field equations of general relativity. And at the center of this approach were tensors. Because tensors have the property of general covariance, the mathe-matical expressions of Einstein's theory would remain unchanged, or invariant, under arbitrary coordinate transformations. The frame of reference could be shifted (i.e., translated) or rotated in space, but the information encoded within a tensor—as well as the relationships between the tensor's individual components—would be unaffected by such shifts. This makes sense, given that the word *covariant* means that separate things, such as the components of a tensor, change to-gether, in ensemble—as opposed to not changing at all, which is what the word *invariant* means.

A four-by-four tensor has sixteen elements or functions. To each of these functions, one can apply the tools of calculus. On a tensor, however, these operations must be performed on functions not indi-vidually but all at once. That required a new kind of calculus, tensor calculus, which is exactly what Ricci and Levi-Civita had invented.

Einstein acknowledged that his focus had truly shifted once he embraced the tools of tensor calculus, noting that "the problem of gravity was thus reduced to a purely mathematical one"[14]—a state-ment he probably never envisioned making when he embarked on his mission to rewrite the laws of gravity. In fact, earlier in his career, Einstein reportedly asserted: "I do not believe in mathematics."[15]

Einstein elaborated on his new focus in an October 1912 letter to the physicist Arnold Sommerfeld: "I occupy myself exclusively with the problem of gravitation and now believe I will overcome all difficulties with the help of a friendly mathematician here. But this one thing is certain: that in all my life I have never labored at all as hard, and that I have become imbued with a great respect for mathematics, the subtle parts of which, in my innocence, I had till now regarded as pure luxury. Compared with this problem, the original theory of relativity is child's play."[16]

In June 1913, Einstein and Grossmann, the aforementioned "friendly mathematician," jointly published the first important paper on general relativity—a first draft that is now referred to as the *Entwurf* (or outline) theory. The paper, "*Entwurf Einer Verallgemeinerten Relativitätstheorie und Einer Theorie der Gravitation*" (Outline of a Generalized Theory of Relativity and a Theory of Gravity), came in two parts: The first section, which focused on physics, was written by Einstein. The second part, focused on mathematics, was written by Grossmann, who was charged with working out the geometry of curved Riemannian space that a gravitational theory required.

This first draft came very close to the "final" version of the equations that Einstein delivered on November 25, 1915. Like the 1915 version, the *Entwurf* paper shows that gravity arises from the curvature of spacetime. Incorporating the absolute differential calculus of Ricci and Levi-Civita, the field equations in this paper are written in terms of tensors, which thereby constitute the fundamental objects of study in general relativity. The elements of a tensor in general relativity encode information about the distance between two points on a manifold that are close to one another, about the curvature of that manifold at a particular point, or about the energy or mass density at a given location in spacetime. The curvature (or geometry) of spacetime, the equations showed, is intimately tied to the distribution of energy and mass.

As written in the paper, the gravitational field tensor occupied the left-hand side of the *Entwurf* field equation, whereas tensors for both matter and energy occupied the right-hand side. In this case, the gravitational field tensor, g_{ij}, was represented by a four-by-four array of sixteen spacetime functions:

$$\begin{bmatrix} g_{11} & g_{12} & g_{13} & g_{14} \\ g_{21} & g_{22} & g_{23} & g_{24} \\ g_{31} & g_{32} & g_{33} & g_{34} \\ g_{41} & g_{42} & g_{43} & g_{44} \end{bmatrix}.$$

However $g_{ij} = g_{ji}$, so the gravitational field is actually characterized by ten distinct spacetime functions rather than by sixteen functions because six of these functions belong to identical pairs.[17] The point of utilizing the calculus of Ricci and Levi-Civita was to make sure that the equations were generally covariant—that they would provide correct physical descriptions regardless of the system of coordinates in which they were expressed. In other words, given that the choice of coordinates is completely arbitrary—selected strictly for the convenience of a person charged with analyzing or describing the situation and not an intrinsic feature of the natural world—changing the coordinates should not have any effect whatsoever on actual physical properties. It's worth noting, moreover, that when coordinates or frames of reference are shifted, each side of a covariant equation will change as well, but both sides will change in exactly the same way.

This is where Einstein and Grossmann ran into a snag. In their first attempt in 1913, they were unable to make their equations fully, or generally, covariant, and ended up abandoning the goal of making them so, even though it had been a fundamental guiding principle of the whole enterprise. As Einstein stated in Part I, the physics portion of the paper: "It seems to follow that the equations sought will be covariant only with respect to a particular group of transformations, which group, however, is as yet unknown to us."[18] The covariance of the *Entwurf* field equations turned out to be severely limited.

Einstein then conjured up physical arguments to justify their failure and explain why general covariance was impossible. He and Grossmann mistakenly believed that a generally covariant theory would not yield correct results at the "weak field" or so-called Newtonian limit, thereby falling short of the goal that Einstein intended to achieve, which was to create a theory that would reduce to Newton's gravitational law in situations where gravity was feeble. He offered other explanations to justify that decision, including a conviction that the conservation of energy and momentum required the covariance of the field equations to be restricted rather than general.[19] Einstein also believed that "a law of gravitation invariant with respect to arbitrary transformations of coordinates was inconsistent with the principle of causality." He later recognized that the explanations he had offered were wrong—"errors of thought which cost me two years of excessively hard work."[20]

The *Entwurf* theory had another shortcoming: It did not correctly predict Mercury's perihelion shift, which had been a key motivating factor for creating a new gravitational theory in the first place.

Using the *Entwurf* equation, Einstein and Michele Besso arrived at an advance of the perihelion of an extra 18 arc-seconds per century as compared to the value obtained from Newton's laws. However, astronomers had indicated that the excess should be on the order of 43 arc-seconds per century—a significant disparity.

Einstein and Grossmann had come close in the *Entwurf* paper— "within a hair's breadth of the generally covariant field equations of the final theory," as science historian John Norton put it.[21] But they had failed to make their equations fully covariant—in part because that objective had proved difficult to achieve and also because, after running into difficulties, they had convinced themselves it was unnecessary to do so. "Nothing is easier for a first-rate mind than to form plausible arguments that what it cannot do cannot be done," the philosophers and science historians John Earman and Clark Glymour would later remark.[22] The decision to abandon the requirement for generally covariant equations was an unfortunate one, given that the condition had long been a guiding principle for the gravitational theory that Einstein had sought, as well as the main motivation for adopting the mathematical framework of Ricci and Levi-Civita.

It would take Einstein more than two years, along with a tremendous effort, to traverse the remaining "hair's breadth" Norton referred to. Einstein would complete the final spacetime interval of his journey without much additional input from Grossmann. Indeed, their collaboration ended up pointing them in the wrong direction. The pair would coauthor just one more paper, a 1914 collaborative effort in which they claimed to have proven that a fully covariant theory was not possible.[23]

In that same year, 1914, Einstein moved to Berlin, accepting an invitation from the physicist Max Planck to join the Prussian Academy of Sciences, where he became director of the Academy's soon-to-be-established Kaiser Wilhelm Institute for Physics. There, Einstein found himself without a steady mathematical partner, yet he still needed outside help in gaining mastery over tensor mathematics and covariance principles, wherein lay the biggest hurdles he faced. He would soon work harder than he had ever worked in his life, pushing himself to the brink, without knowing—until the very end—where his labors would take him, or whether he was even heading in the right direction.

3

The Magnus Opus

After taking leave of Zurich and Marcel Grossmann and moving to Berlin in 1914, Albert Einstein was at an impasse, feeling that he was close yet still struggling to find a pathway to the final theory. He worked feverishly on his own, while also gaining some vital succor along the way. Some of that critical assistance came from Tullio Levi-Civita, with whom Einstein had struck up a fruitful correspondence. Their conversations began in early 1915, most likely in March, after Levi-Civita noticed some flaws in a general relativity paper Einstein had published in November 1914—in particular, a tensorial expression in the left-hand side of an earlier version of his field equations.

Einstein issued a polite reply in his defense, stating that "upon thorough consideration I do believe I can uphold my proof."[1] Einstein did not cede his ground easily. He continued to debate the Italian mathematician for at least another two months, responding to Levi-Civita's objections with his own rebuttals, which were invariably refuted by the latter. In a May 5, 1915, letter Einstein finally admitted that the proof he had defended so vigorously over the past couple of months was "incomplete."[2] One benefit of this grueling, back-and-forth process is that his understanding of tensor calculus surely must have increased.

Despite his initial resistance, Einstein expressed his gratitude for these exchanges in a letter to another friend: "Only one, Levi-Civita in Padua, has probably grasped the main point completely, because he is familiar with the mathematics used. Corresponding with him is unusually interesting; it is currently my favorite pastime."[3] Einstein clearly respected Levi-Civita's mathematical abilities, telling him: "I admire the elegance of your method of computation; it must be nice to ride through these fields upon the horse of true mathematics while the likes of us have to make our way laboriously on foot."[4]

The next big step in Einstein's education came after he accepted an invitation from David Hilbert, widely regarded as the world's leading mathematician, to come to Göttingen, which was then considered the world's center of mathematical research. In late June and early July of 1915, Einstein delivered six two-hour lectures on general relativity.

At that time, he was still adhering to the restricted covariance of the *Entwurf* equations. In letters to his colleagues, Einstein reported his "great joy...in convincing Hilbert and [Felix] Klein completely" of his arguments—or so he thought. Einstein expressed his enthusiasm about Hilbert in particular, calling him "an important figure,"[5] which many would consider an understatement. Yet it also seems clear that Einstein's discussions with virtuoso mathematicians like Hilbert and Klein eventually convinced him that the lack of general covariance in his (and Grossmann's) field equations was a serious deficiency. Because of that—and the fact that the *Entwurf* equations did not, at least in his prior attempt with Michele Besso, yield the correct value for the precession of Mercury's perihelion—he finally acknowledged that his work was unfinished.[6]

A couple of months after his Göttingen lectures, Einstein learned that Hilbert had set out to rewrite the equations of the *Entwurf* theory, and of gravitation in general, taking an axiomatic approach, which is to say starting solely from mathematical principles. As Hilbert stated in his paper, "The Foundations of Physics (First Communication)," which was presented at a talk in Göttingen on November 20, 1915, and published in early 1916: "I would like to develop, essentially from two simple axioms, a new system of basic equations of physics of ideal beauty."[7]

Hilbert's approach was grounded in the field of variational calculus, also known as the calculus of variations. Although his exact motivation is difficult to discern, it seems likely that Hilbert saw a much more direct method of deriving the field equations—by harnessing the sheer power of his mathematical expertise—than the multiyear approach taken by Einstein, which relied to some extent on trial and error. In an oft-repeated quote, Hilbert famously said that "physics is much too hard for physicists."[8]

Einstein was almost certainly displeased to hear that he had acquired so formidable a rival. Just as Einstein had initially characterized Minkowski's foray into special relativity as "superfluous learnedness," he was also suspicious of Hilbert's approach, complaining in a letter to Hermann Weyl (a former student of Hilbert) that basing the equations of physics purely on mathematics, without relying on any input from experiments, "appears to be childish, just like an infant who is unaware of the pitfalls of the real world."[9] That said, Einstein recognized Hilbert's unquestionable abilities, which must have motivated him to hasten his efforts to complete his own version of the gravitational field equations. With the pressure on, things fell into place for Einstein in the month of November, during which he produced four successive papers—one coming out each week, and each reflecting the steady evolution of his thinking, likely abetted by his exchanges with Hilbert.

During October and November of 1915, he and Hilbert corresponded frequently—each keeping the other apprised of their progress. And in the final weeks, between November 7 and 25, Einstein corresponded only with Hilbert and no one else.[10] However he felt about his rival, Hilbert's influence on Einstein was profound: "It is quite clear from the November correspondence [between Einstein and Hilbert] (and from recently discovered letters of Max Born to Hilbert of the fall of 1915) . . . that Hilbert's competitive influence was crucial for Einstein's acceptance of general covariance—in spite of his long-time reservations and doubts," the physicist Ivan Todorov commented.[11]

A milestone was achieved in his November 18 paper, when Einstein announced that the perihelion motion of Mercury could be explained using his latest version of his general relativity theory.[12] "For a few days, I was beside myself with joyous excitement," Einstein recounted afterward.[13] In a letter dated November 19, Hilbert congratulated him on this achievement and on the speed of his Mercury calculation. But, of course, Einstein had not started from scratch on this problem; he was merely revisiting the calculation he and Besso had carried out in 1913—this time arriving at an answer of 43 arcseconds per century for the change in orientation of Mercury's elliptical orbit, which closely agreed with the observed value. And exactly one week later, on November 25, 1915, Einstein came out with his final draft of "The Field Equations of Gravitation."

Hilbert had submitted his paper, "The Foundations of Physics," on November 20, five days earlier, which in its published form had almost identical field equations. The historians Jürgen Renn and John Stachel called Hilbert's contributions to general relativity "a unique and independent achievement.... It appears that Hilbert indeed had found an independent 'royal road' to general relativity and beyond."[14]

Hilbert did in fact believe that his "Foundations" paper went "beyond" general relativity. In addition to reinforcing Einstein's new views about space, time, and motion, Hilbert wrote, "I am also convinced that through the basic equations established here, the most intimate, hitherto hidden processes in the interior of atoms will receive an explanation; and in particular that generally a reduction of all physical constants to mathematical constants must be possible—whereby the possibility approaches that physics in principle becomes a science of the type of geometry."[15]

Nevertheless, the question of priority—of who was the first to get the equations of general relativity right—remains rather murky, given that some science historians have maintained that Hilbert made significant changes to that paper after it was submitted but prior to its publication on March 30, 1916.[16] The amount of help that Einstein got from Hilbert, and Hilbert from Einstein, is also a bit unclear.

"Because Einstein and Hilbert exchanged notes with each other during their four weeks of intense activity in the fall of 1915, [their] respective contributions have been somewhat difficult to entangle," wrote the astronomer Martin Harwit, former director of the National Air and Space Museum. Yet Harwit believes that both parties benefited from these interactions, stating that "Hilbert's approach considerably influenced Einstein," while also noting that "Hilbert always acknowledged that it was Einstein's physical insights that had aroused his own interests in finding such a set of equations."[17]

Even though the issue is not completely settled, it now seems reasonable to talk about "Einstein's general theory of relativity," given that Einstein had worked out the physical basis of this theory mostly on his own, while at the same time referring to the "Einstein–Hilbert field equations" as a kind of joint endeavor.

Here's how the physicist Kip Thorne viewed their respective contributions to the field equations: "Hilbert had carried out the last few mathematical steps to its discovery independently and almost simultaneously with Einstein, but Einstein was responsible for essentially everything that preceded those steps."[18] In the final sentence of their paper, "Einstein and Hilbert," the historians John Earman and Clark Glymour provided a satisfactory conclusion to the priority issue: "Hilbert's recognition of Einstein's undeniable authorship of both the general framework and the central ideas of the theory may well account for the fact that Hilbert never claimed credit for the general theory of relativity."[19]

Unlike the priority dispute between Newton and Leibniz over the invention of calculus, which was never resolved during their lifetimes, Einstein and Hilbert were able to put aside their differences and move on from this matter. This discussion, accordingly, shall do the same, focusing instead on the actual substance of Einstein's and Hilbert's parallel contributions.

The eight years of arduous labor, following Einstein's "happiest thought" of 1907—when he was struck by the critical importance of the equivalence principle—can be boiled down to a single equation with just a handful of terms, which appeared in his November 25, 1915, paper (that was published on December 2, exactly one week later):

$$R_{ij} - \frac{1}{2} R g_{ij} = T_{ij}.$$

This equation can be rewritten in an even simpler form by calling the entire left-hand side of the equation,

$$R_{ij} - \frac{1}{2} R g_{ij},$$

the Einstein tensor, G_{ij}, so that the whole equation then reduces to

$$G_{ij} = T_{ij}.$$

This appears deceptively simple. We must remember (as discussed in the previous chapter) that the subscripts, i and j, are actually variables that can assume four different values (0, 1, 2, or 3)—each associated with a different direction, or dimension, in spacetime (or degree of freedom, as physicists and mathematicians sometimes call it). The 0 corresponds to the time direction, and 1, 2, and 3 correspond to the x, y, and z spatial directions, respectively. First, owing to the fact that i and j can take on four values, the components of the above tensors are functions of four variables, which makes the above equation, consisting of just two terms, much more complicated than it appears at first glance. Furthermore,

$$R_{ij} - \frac{1}{2}Rg_{ij} = T_{ij}$$

is not just a single equation but is, in fact, sixteen equations when all possible combinations of i's and j's are entered in, even if only ten of those equations are independent of each other.

We'll now examine the terms of this famous equation, one by one, before discussing the full weight of what happens when they are put together in the proper way.

Let's start with R_{ij}, the Ricci (curvature) tensor, which is derived from the Riemann curvature tensor. Ricci introduced this tensor in the late 1800s, well before the advent of general relativity and before anyone considered that it might have anything to do with gravity. Using a process that mathematicians call contraction, Ricci was able to break down the Riemann tensor—classified as rank 4, owing to the four subscripts—into its constituent parts, one of which is a shrunken-down (rank 2) version of the Riemann tensor called the Ricci tensor. It has two indices or subscripts instead of four, though its elements are still functions of four variables. In four-dimensional space, or spacetime, the Ricci tensor does not contain all of the curvature information of the Riemann tensor, but, fortuitously, it contains a key portion of the information that Einstein needed. Plus, it had the advantage of being much easier to work with than the full curvature tensor.

In the 1913 *Entwurf* paper, Grossmann noted that the Ricci tensor seemed like the right choice for the gravitation tensor, yet he concluded, erroneously, that the tensor would not reproduce Newton's theory of gravity at the limit of extremely weak gravitational fields. Einstein concurred with that assessment, and the resulting field equations—as of June 1913—did not incorporate the Ricci tensor.[20] It was not until November 4, 1915—after nearly two and a half years of hard work, involving numerous twists and turns, as well as some dead ends—that Einstein first used the Ricci tensor to represent gravity in his field equations.[21] That, of course, was a critical step that enabled him to reach the equations' final form three weeks later.

The next term in the field equations is R, the scalar curvature tensor, which is also known as the Ricci scalar. It's called a scalar because at any point in spacetime, the tensor assigns a single number. The scalar curvature is the simplest curvature property, or invariant, of a Riemannian manifold, and it is the generalization of Gauss's two-dimensional intrinsic curvature to an arbitrary number of dimensions.

The scalar curvature tensor is derived from—and is, in fact, a contraction of—the aforementioned Ricci tensor. Consequently, it carries less curvature information than the Ricci tensor, which itself provides only a small amount of curvature information as compared to the Riemann tensor. How the scalar curvature tensor contracts the Ricci tensor and can be reduced to a single number is rather simple to explain. Let's think of the Ricci tensor as a four-by-four array consisting of sixteen functions. At any given point in four-dimensional spacetime, if you plug in the specific coordinates, each of those functions will give you a number. In a suitable, so-called normal coordinate system, the scalar curvature (in a Lorentz metric) can be determined by adding the three spatial components along the diagonal (going from the upper left-hand corner to the lower right-hand corner of the tensor) and subtracting the time component. But, again, this simple procedure only works in a special coordinate system—one that could offer a convenient way of describing our universe.

The metric tensor g_{ij} describes the geometrical properties of spacetime—including its curvature—at any and every point. Identifying the components of this tensor is, in essence, the name of the game in general relativity because the gravitational effect stems entirely from the curving or bending of spacetime.

That pretty much covers the left-hand side of the equation, which is encompassed within the catch-all term G_{ij}, the Einstein tensor that represents the curvature of spacetime.

Remarkably, although the Einstein tensor, $R_{ij} - \frac{1}{2}Rg_{ij}$, did not make its first appearance in general relativity until November 1915, that very same tensorial expression had come up many years earlier in a mathematical context that had nothing to do with Einstein or gravity.

In papers written separately by three mathematicians—Aurel Voss in 1880, Ricci in 1898, and Luigi Bianchi in 1902—constructs later called the contracted Bianchi identities were independently derived.[22] The identities relate to the so-called divergence of the Ricci tensor.

Divergence, in nontechnical language, has to do with how much stuff—be it electric charge, matter, energy, or even water—is flowing into or out of a particular space. Voss, Ricci, and Bianchi were thinking in terms of vectors, trying to see whether the overall flux of vectors within a particular region was pointing in or out (or not at all). In a space that has zero divergence, there is no net flow in any direction. In other words, energy (or any other commodity you might care to look at) is conserved.

These mathematicians calculated the divergence of the Ricci tensor, R_{ij}, and it turned out to be exactly equal to the divergence of $\frac{1}{2} R g_{ij}$. That equivalence, in turn, implied that the divergence of $R_{ij} - \frac{1}{2} R g_{ij}$ (otherwise known as G_{ij}) has to be zero. That is another way of saying that the Einstein tensor, G_{ij}, satisfies the law of energy conservation because, with zero divergence, no net energy can leave or enter the system.

This was an essential property to know about the Einstein tensor— and about the November 25, 1915, version of the field equations as a whole. But these earlier papers by Voss, Ricci, and Bianchi were not known to Einstein at the time, nor had they come to the attention of his contemporaries like Hilbert and Klein.[23] Part of the reason for that may have been due to the fact that the Bianchi identities—and the contracted Bianchi identities derived from them—did not appear in the 1910 German edition of the book of Bianchi's collected lectures.[24] For whatever reason, Einstein did not realize that the tensor soon to be named after him had already been derived by other mathematicians, and it took him several long years to come up with that same formulation on his own.

Given that mathematicians had shown that G_{ij} is divergence-free, and given that Einstein later showed that $G_{ij} = T_{ij}$, we can thus conclude that T_{ij} must be divergence-free as well. Another way of putting that is that energy conservation prevails on both sides of the equation. (The fact that T_{ij} satisfies the conservation law was already known from classical mechanics, although that statement was not expressed in terms of tensors.) Of course, we have not yet explained what T_{ij}, the right-hand side of this famous equation, is all about, so this might be an appropriate time to do so.

For starters, T_{ij} is called the stress-energy or energy-momentum tensor. Encapsulated within the components of this tensor is a bounty of information regarding the density, flow, and momentum of matter and energy—in other words, how matter and energy (of a non-gravitational form) are distributed and moving throughout spacetime. Density, in this case, can be expressed in the usual way, such as grams per cubic centimeter. This quantity would be zero in a region devoid of matter, whereas it could be staggeringly high in a region (such as the interior of a neutron star) that is jam-packed with matter.

Astronomers, in principle, can gather enough experimental data to provide us with a pretty good handle on T_{ij}. In a vacuum region, that is an especially easy task because T_{ij} then equals zero. (In that case, the so-called vacuum field equation reduces to $G_{ij} = 0$.)

The major unknown here is the metric tensor, g_{ij}, and the equations provide a framework for figuring out what it is, thereby shedding light on the curvature question that can, in turn, answer any questions you might have about gravity. If you can solve the Einstein (or Einstein–Hilbert) equations—which, as will be discussed later, is a very demanding task—you might conceivably determine the components of the metric tensor. From there, you could get the Riemann curvature tensor, encompassing all varieties of curvature information. The Ricci tensor can be derived from that, and the scalar curvature tensor, in turn, can be derived from the Ricci tensor. And thus, all the pieces can, in principle, fall into place.

In reality, however, things are a bit more complicated. It turns out that the Einstein equations, alone, are not enough to completely determine g_{ij} or to specify every aspect of curvature, even though Einstein might have thought that was the case early on. In any dynamical situation, where things are changing in time, one must also know what the initial conditions are, as well as the so-called boundary conditions. A toy model of both sets of conditions would be a rubber band stretched between two pegs. The position of those pegs gives you a pretty firm indication of what the boundary conditions are in this case. If you then pull the band out sideways to a certain point—just before releasing it—you will also know what the initial conditions are. With that knowledge, you can accurately predict the vibrations of the rubber band for some time into the future (although maybe not forever). As for the universe, the initial conditions—and origins of it all—are still a huge mystery, as is the nature of the boundary itself. We don't yet know whether the universe has an actual "edge" and, if not, how it behaves asymptotically, as infinity is approached.

These issues will, to some extent, be delved into in future chapters. But for now, let's return our gaze to the subject at hand, the field equations that Einstein obtained after roughly a decade-long struggle:

$$R_{ij} - \frac{1}{2}Rg_{ij} = T_{ij},$$

or, again, in their abbreviated form,

$$G_{ij} = T_{ij}.$$

Newton's law of gravitation, expressed in the language of differential calculus, called the Poisson equation for gravity, has a very similar form. That, of course, is no accident, given that Einstein was, from the very beginning, trying to generalize Newton's law and preserve the successes of the earlier theory. The left-hand side of the Newton/Poisson equation is basically the second derivative of a function related to the gravitational potential energy, while the right-hand side is a function that tells you the mass density anywhere and everywhere in space.

In some ways, it's not so different from $G_{ij} = T_{ij}$, though there is a rather important distinction: Newton's law is represented by a single differential equation composed of so-called scalar functions. If you plug in, say, the x, y, and z coordinates of a point in space, a scalar function will spit out a single number.

At what's called the Newtonian limit—where gravity is extremely weak and spacetime essentially flat—Einstein's formulation reduces to the differential version of Newton's law described above. But in everything but that one special case, the richness and complexity of tensorial notation is needed to describe gravity in four-dimensional spacetime. That should come as no surprise at this juncture given that we've seen how hard Einstein labored to familiarize himself with the language of tensors before he could even begin to write the gravitational field equations.

As with the initial 1913 *Entwurf* paper, the left-hand side of his November 25, 1915, equation again relates to the curvature of spacetime. This property cannot be ascertained through observations, since it's not possible to step outside our spacetime in order to measure its curvature. Spacetime curvature can only be determined intrinsically, by drawing upon geometric principles—just as Eratosthenes (circa 200 BCE), who had no access to rocket ships or satellites, instead relied on good old-fashioned geometry and other logical arguments to work out the curvature of Earth. And he managed to do so with surprising accuracy.

The right-hand side of the equation (returning, once again, to the early twentieth century) relates to the movement and distribution of matter and energy—quantities for which empirical evidence can, in principle, be obtained. Perhaps the biggest change from 1913 to 1915 was that the latter equations maintain general (rather than restricted) covariance—in keeping with the equivalence principle, which had been Einstein's goal, and guiding tenet, all along.

The expression $G_{ij} = T_{ij}$ is not just shorthand for ten independent and interconnected equations. They happen to be equations of a very particular sort—nonlinear, second-order, partial differential equations of four independent variables—that normally cannot be solved in full generality. Their solutions can only be obtained approximately or in cases when simplifying assumptions are made. (The latter point, involving special, idealized cases that have subsequently been solved, will be taken up in the next chapter.)

These equations, which tie together two things previously considered disconnected, substantiate the premise in which Einstein had believed for many years—namely that the curvature of spacetime, or gravity, is largely determined by the distribution of mass and energy, and the converse is true as well. The left-hand side of the equations represents the curvature of spacetime. The right-hand side represents mass and energy. Expressed in other terms, Einstein's equations tell us that the thing we've been calling gravity is not a force at all. It's merely a consequence of the curvature of spacetime. And that curvature, in turn, is determined—to a great extent—by the distribution of matter and energy in its initial state (at "time zero," you might say), how things are changing and moving in time (i.e., the momentum of matter and energy), and the topology (or overall shape) of the universe. In other words, a lot of factors go into the determination of spacetime curvature.

Nonlinear effects can arise, in both mathematics and physics, when the main variables influence each other as profoundly as they do in gravity. And that's just how nonlinearity comes into general relativity, complicating matters greatly. Massive objects curve spacetime, and that generates the thing we call gravity. But gravity is, itself, a form of energy, which Einstein taught us is interchangeable with mass ($E = mc^2$). The presence of that energy can cause additional curvature of spacetime, thereby creating even more gravity—or, as it's sometimes called, the gravity of gravity. And there is even a more subtle effect: While matter can certainly drive geometry, geometry can also interact with itself—even in a spacetime wholly lacking in matter—and curvature will evolve from there, according to this nonlinear interaction.

It's clear then that in the gravitational arena, as was the case in special relativity, space and time are not merely passive stages upon which physical transactions take place. Instead, they are active players in the physical world—constantly changing and distorting in response to the shifting distributions of matter and energy.

Gravity curves spacetime and thereby changes its geometry. But objects "falling" freely under the influence of gravity are not responding to an external force. They are merely taking the shortest, most direct path available to them through curved spacetime—what would, to them, seem like a straight, downhill journey—and that's true even if the trajectory they are tracing out happens to be curved.

Einstein explained the idea this way to his nine-year-old son, who'd asked him why he had suddenly become so famous: "When a blind beetle crawls over the surface of a curved branch, it doesn't notice that the track it has covered is indeed curved," Einstein replied. "I was lucky enough to notice what the beetle didn't notice."[25]

Reaching that insight, as discussed, had been a grueling ordeal. "In the light of knowledge attained, the happy achievement seems almost a matter of course, and any intelligent student can grasp it without too much trouble," Einstein wrote in a November 1916 paper, published one year after his eventual triumph. "But the years of anxious searching in the dark, with their intense longing, their alternations of confidence and exhaustion and the final emergence into the light, only those who have experienced it can understand that."[26]

In his major paper, "The Foundation of the General Theory of Relativity," which was submitted on March 20, 1916, and published in *Annalen der Physik* on May 11, 1916, Einstein did thank his friend

Grossmann, who "helped me in my search for the field equations of gravitation." However, Einstein was not always generous in his acknowledgement of Grossmann's contributions to this effort, commenting on one occasion that Grossmann "only helped in guiding me through the mathematical literature but contributed nothing of substance to the results."[27] Grossmann, of course, did quite a bit more, writing the mathematical portion of the *Entwurf* paper, which is significant because it contained the precursors of the final field equations. Grossmann also deserves credit for introducing the term *tensor*—in place of Ricci and Levi-Civita's terminology, *covariant* and *contravariant systems*—as well as for changing the notation (in terms of subscripts and superscripts) used to characterize tensors in order to make them more useful in both mathematics and physics.

Grossmann surely played more than a cameo role in these proceedings, yet he never tried to grab the limelight. Instead, he applauded his friend's efforts without reservation and without laying claims to being any sort of co-discoverer of general relativity. "To a person who witnessed Einstein's first laborious attempt in 1912 and 1913, as the composers of these lines did, they must appear like the ascent of an inaccessible mountain in the dark of the night, without path or trail, without foothold or direction," Grossmann wrote. "Experience and deduction provided only few and insecure footholds. All the higher we have to value this intellectual deed."[28]

In addition to Grossmann, Einstein also acknowledged the work of Gauss, Riemann, Christoffel, Ricci, and Levi-Civita. One person he ignored was David Hilbert, with whom he was fiercely vying for priority and who, Einstein claimed at one point, was trying to "nostrify" (or plagiarize) Einstein's theory.

Hilbert's achievement, however, came through an entirely different approach, and it didn't take long for Einstein and him to settle their differences. Moreover, the method Hilbert used for deriving the field equations is now considered almost as important as the result he achieved. For he was the first person to successfully derive the field equations of general relativity from the principle of least action, also called the action principle, which is a direct and efficient approach that has become almost ubiquitous throughout contemporary physics.

The origins of this principle can be traced back to Euclid or Archimedes, both of whom recognized that, in a plane, the shortest distance between two points is a straight line. The French mathematician Pierre de Fermat developed this idea further in the mid-1600s, proposing what's now called Fermat's principle—namely that among all possible paths that light could take in moving from one point to another, it always follows the path that requires the least amount of time.

It took physicists a while to work out a corresponding principle that governs the equations of motion for matter particles, which unlike light, are not constrained to travel at a constant speed. That required the introduction of a more general concept called the action, which can represent whatever quantity it is—distance, time, curvature, or a function of some sort—that you are hoping to minimize, or in some cases maximize. In either case, one is seeking to find an extreme of some sort, either low or high. The mathematicians

Gottfried Leibniz, Leonhard Euler, and Pierre Louis Maupertuis all contributed to the principle of least action during the first half of the eighteenth century, but in the latter part of that century, the mathematician Joseph-Louis Lagrange recast the action principle into an even more general and broadly applicable form, maintaining a firm conviction that his methods placed the subject of mechanics—which concerns the motion of physical objects—within the realm of pure mathematics.

In Lagrange's framework, the action, which is represented by S, is the integral of a function called the Lagrangian whose minima and maxima are determined through the calculus of variations, which Euler, and then Lagrange, largely developed. Just as Newton had to invent calculus, a new field of mathematics, in order to express his second law in precise mathematical language, Lagrange needed the calculus of variations in order to go from just having an expression of the action to actually figuring out the related equations of motion.

This area of mathematics revolves around finding suitable values for parameters that can, in turn, yield solutions with critical values, including minimal or maximal values. The calculus of variations can address, for instance, so-called isoperimetric problems of the sort solved by the Greek mathematician Zenodorus in the second century BCE, such as finding the maximum area that can be enclosed by a loop. The answer in this case happens to be a circle.[29] But the calculus of variations can also address more complicated, higher-dimensional problems such as finding a solid that has the least surface area for a given volume—or finding the maxima and minima of more generalized functions.

As the physicist Cumrun Vafa explained, "the calculus of variations... is more complicated than finding minima of a function of a finite number of variables because there are infinitely many paths that connect two points in space. So in a sense it is equivalent to finding the minimum of a function (the action) of infinitely many variables (constituting the space of all paths). Physicists could [then] use the calculus of variations... to pick out the path of minimum possible length."[30] Once you have identified the path that a particle will follow and know its displacement over time, you can determine its velocity and acceleration and from there work out the equations of motion.

For the purposes of mechanics, Lagrange defined the Lagrangian, L, as the particle's kinetic energy, K, minus its potential energy, V, as the particle moves over time from one point in space to another. The action, being an integral, is just the sum of the Lagrangians at every point of the particle's journey through space and time.

If the Lagrangian, L, is chosen to be $K - V$, then the action, S, is equal to the integral of L over time:

$$S = \int L \, dt = \int (K - V) \, dt.$$

There are infinite ways that a particle can move from A to B. Finding the minimal value of S is equivalent to finding the path of least energy, and that, Lagrange presumed, is the path a particle would inevitably take. Working from that basis, one could generate the equations of motion for a particle and thereby derive Newton's famed second law of motion: force equals mass times acceleration, or $F = ma$. The approach Newton used in describing a particle's path relies heavily on differential calculus, which focuses on how physical quantities, such as velocity, change as a particle moves from one moment, and one point in space, to the next. The particle's overall trajectory is determined from repeated differential assessments—tiny, iterative adjustments—of this sort, carried out each step of the way. In relying on the action principle, one takes a different and more holistic approach, using the integration of a single quantity—the action—to show why a particle is destined to follow one path rather than another.

Hilbert, as already indicated, opted for the second approach, which is in many ways simpler and more straightforward. He, of course, had no interest in reproducing Newton's second law some 250 years after Newton had already established it. Hilbert, focused as he was on general relativity and its curved spacetime, therefore needed to pick a different action to be minimized—or extremized—than, say, $L - V$. And it wasn't simply a matter of minimizing time, as Fermat did in describing the pathway of a light beam, or of just minimizing length by finding the shortest distance between two points on a plane. The whole objective of general relativity is to figure out how a given distribution of matter and energy affects, and literally curves, spacetime. Therefore, the action that Hilbert chose to extremize would have to involve the curvature of spacetime, but there was still the matter of picking the right expression of curvature to use.

In making this choice, Hilbert drew on his deep expertise in the theory of invariants—a field that essentially started with work in the 1840s by the mathematician Arthur Cayley, though Hilbert made important contributions in this area some decades later. An invariant, again, is a property inherent to a mathematical object that does not change, even after repeated transformations are performed on that object. In the context of a gravitational theory, Hilbert knew there were just two invariants that varied linearly with the Riemann curvature tensor—meaning that if the value of one changed, the other would change a proportional amount. One invariant is the aforementioned scalar curvature tensor, R, and the other is a constant function that has the same value everywhere in space. The scalar curvature tensor turned out to be the correct choice for Hilbert, as well as the simplest invariant he could use for this purpose.

Hilbert posited that, in this case, the Lagrangian, L, should be equal to R. The action then is basically the integral of the scalar curvature over space and time, and from there he used the calculus of variations to derive the same field equations that Einstein obtained by other, more roundabout (and also more grueling) methods that were familiar to him and that had served him well in the past. As the physicist David Garfinkle pointed out, "you first have to be convinced that the equations of physics, and the laws of nature, stem from an action principle" to consider following Hilbert's route.[31] That may seem obvious today, Garfinkle argued, but more than 100 years ago, when Hilbert turned his attention and talents toward general relativity and put his faith in the action principle, that conviction was not widely embraced.

One person who shared Hilbert's viewpoint was the mathematician Emmy Noether. In 1915, Hilbert and Klein invited her to Göttingen to investigate, among other things, how the notion of energy conservation fit with the new equations of gravity. In particular, Noether looked into an assertion made by Hilbert that conservation of energy had a different status in generally covariant theories than it did in theories that are not generally covariant.[32]

Before describing what Noether did to confirm Hilbert's claim—and to show why energy conservation in general relativity works differently than in previous physical theories—it is worth emphasizing this remarkable fact: Of all the people whom Hilbert (who was then probably the greatest living mathematician alive) and Klein (a world-renowned mathematician in his own right) might have brought to Göttingen to study this issue, they chose her. Einstein (probably the greatest physicist of his era) also welcomed Noether's assistance on a problem with which he, Hilbert, and Klein were still grappling. He told Hilbert in a May 30, 1916, letter "that I understand everything in your article except the energy theorem. Of course," Einstein added, "it would be sufficient if you asked Miss Noether to clarify this for me." This letter and a prior note from Hilbert to Einstein showed that both scientists recognized Noether's expertise in this area—a confidence, it turns out, that was well founded.[33]

What made their choice particularly surprising was that, at the time, Noether did not even have a job in mathematics and she'd barely been able to get an education in the field. When she was of college age, around the year 1900, German universities did not admit women at all, so Noether was forced to audit classes instead. Those policies, which had kept women from the halls of academia, were relaxed a few years later. After educating herself, Noether was admitted in 1904 to a graduate program in mathematics at the University of Erlangen, where she received her PhD just three years later. Noether worked at Erlangen for the next eight years without pay or an official position. When she finally came to Göttingen in 1915, she served as a lecturer, again receiving no pay until her promotion to untenured associate mathematics professor in 1922.

However sparing the academic world had been to her, Noether, nevertheless, was not sparing in her contributions to mathematics—especially in abstract algebra, her main area of study—and to physics. Noether addressed Hilbert's contention, that energy conservation works differently in a theory that is generally covariant, and confirmed it in a 1918 paper called "Invariant Variational Problems," which was dedicated to Klein, as she put it, to celebrate the fiftieth anniversary of his doctorate.[34]

Noether's paper contained two important theorems. We'll start with the second theorem, the lesser known of the two, because that relates most closely to the problem of energy conservation. In the course of her examination of that issue, she was led to what's now called her first theorem, or simply Noether's theorem—an achievement that has been, to put it mildly, hugely consequential.

Noether's second theorem basically showed that energy in general relativity is conserved only if you take a global view—which means looking at a system from a great distance (i.e., infinity). Normal energy conservation—where an observer happens to be in, or near, the system in question—does not hold in general relativity, as it had in previous physical theories such as electromagnetism, because one has to take into account not only the kinetic and potential energy of matter but also the energy stored in the gravitational field itself. The value of the latter contribution, moreover, will vary for nearby observers in different locations, which means there is no single "right" answer. Furthermore, any such measurement would be complicated by the fact that energy is continuously exchanged between matter and the gravitational field. As a result, energy is only conserved when one considers, from afar, the total energy—due to both matter and gravity—confined within an isolated region of space.

Here's a simple way of picturing what she discovered: Suppose you have a bucket of water sitting in an absolutely parched desert on a hot sunny day. If all you look at is the bucket, then the water contained within it will not be conserved because some of it will inevitably evaporate. But if you look at a big enough volume of space—one that encompasses the bucket, the water in it, and all the evaporated water in the air surrounding it—then the total water (of both liquid and gaseous forms) will be conserved. In this same way, when one views a region of space from a great distance, the total energy held in the form of matter and the gravitational field is also conserved.[35]

Noether's finding thus resolved the questions raised by Hilbert and others regarding how energy conservation worked in the field equations of general relativity. This notion had already been indicated by the contracted Bianchi identities, of which she was not aware, but that turned out to be just a special case of the much broader theorem that Noether proved.

The action principle and calculus of variations were front and center in the proof of Noether's second theorem, and they are absolutely crucial to the workings of her first, and by far more famous, theorem.

That theorem basically says that for every continuous symmetry in nature, along with an accompanying action principle, there is an associated conservation law.

Symmetry here refers to an operation you can do to an object or system that leaves it unchanged. If you rotate a square around its center by 90 degrees, for instance, it looks the same, but it doesn't look the same if you rotate it by 45 degrees or 17 degrees or anything other than a multiple of 90. That's an example of a discrete symmetry, whereas the rotation of a circle around its center is continuously symmetric: No matter what degree, or fraction of a degree, you choose to rotate it, the circle will end up looking the same.

Here's how the action principle comes into play: The action is a number you can find by taking the integral of the trajectory of some moving object, say, a cannonball. If you shoot a cannonball and chart its trajectory over space and time, you can compute the action. If you do everything the same 10 seconds later, shooting another cannonball of equal size and weight, you should—under identical conditions (wind speed and direction, etc.)—get the same trajectory and hence compute the same action. Expressed in other words, the symmetry of time translation leaves the action unchanged.

One could just as well shoot the cannon and map out the cannonball's trajectory and then move the cannon one meter to the left on perfectly level ground and repeat everything as before, under the same conditions. Thanks to the symmetry of space translation, the second cannonball will have a trajectory that is identical to the first, and the action, consequently, will be identical as well.

And if, once again, one was to shoot off another cannonball and then do everything the same except for rotating the cannon by some number of degrees from its original position, one would again compute the same action in both instances (if conditions were unchanged). The conserved quantity this time, Noether's first theorem tells us, would be angular momentum.

Noether's theorem tells us not just that there is a conserved quantity associated with a particular symmetry but how to identify it. The conserved quantity associated with space translation is momentum, and the conserved quantity associated with time translation is energy. Prior to Noether, a fair amount of guesswork had gone into the formulation of the laws of mechanics and the conservation principles that went with them, commented the physicist Chris Quigg: "Even something as fundamental as the law of conservation of energy was sort of an empirical regularity. It didn't come from anywhere, but it had been found to be a useful construct. After Noether's Theorem I, we know that energy conservation does come from somewhere that seems rather plausible: the idea that the laws of nature should be independent of time."[36]

Noether's first theorem has had a profound impact on the way physics is done. When speculating about new particles, theorists often postulate a symmetry of nature that they believe should exist but has somehow remained hidden. They would then try to determine the properties of the as-of-yet-unseen particle, or particles, tied to this hypothetical symmetry so that experimentalists know what to look for. Reasoning along these lines led to the discovery of the Higgs boson in 2012, forty-eight years after its existence was first predicted by theorists. As the physicist Ruth Gregory asserted: "It is hard to overstate the importance of Noether's work in modern physics. Her basic insights on symmetry underlie our methods, our theories, and our intuition. The link between symmetry and conservation is how we describe our world."[37]

Despite the enormous influence of Noether's first theorem on physics, it should be emphasized that her result was strictly mathematical—a robust statement that would still be significant even if there hadn't been any physical applications. A key starting point for this powerful theorem was variational calculus. "Theorem I," wrote math historian David Rowe, "precisely characterizes how conserved quantities arise from symmetries in variational systems."[38]

While Einstein had not followed that approach initially, he did venture down this same mathematical route in a paper dated November 26, 1916—one year and a day after he had presented his final version of the field equations in Göttingen. In this paper entitled "Hamilton's Principle and the General Theory of Relativity," Einstein stated that he would—as Hilbert had done before him—derive the equations of general relativity "from a single variational principle." But unlike Hilbert, he added, "I shall make as few assumptions about the constitution of matter as possible. On the other hand, and in contrast to my own very recent treatment of the subject matter, the choice of a system of coordinates shall remain completely free."[39]

Over the months and years that followed, the equations that Einstein labored so hard to discover would pass a series of tests: not just the prediction of the movements of Mercury, but also the famous bending of light by the sun. But a bigger question loomed: Now that we had some confidence in the equations, what more could be done with them?

Quite a bit, as it turned out. The physicist Hanoch Gutfreund called Einstein's work in his November 25, 1915, paper "the source and basis for everything we know today in modern cosmology—how the universe started, the Big Bang theory, the expanding universe, black holes, gravitational waves. Everything follows from that paper." Not only from that paper, but from a single equation on page 33 of the published version, he added. "Everything we know follows from that equation."[40] We can use to it to determine the properties of a given spacetime—including the one we call home—and the physical phenomena to which it gives rise.

Fair enough. But the equation in question is really just a starting point rather than an endpoint. How to go from there to "everything we know" turns out to be no easy feat.

4

A Most Singular Solution

Deriving the field equations of general relativity was a monumental feat that is still celebrated today. But even at his moment of triumph, Albert Einstein was not sure it would be possible to find exact solutions to the coupled nonlinear partial differential equations he had worked so hard to come up with, and he was skeptical of some of the solutions that others had put forth. When he had calculated the shift in Mercury's orbit and later predicted how the sun would bend light from a distant star, he had sought, and ultimately found, approximate—rather than exact—solutions to his equations. "Although his work had revolutionary implications," wrote the physicist Brandon Carter, "Einstein's instincts tended to be rather conservative."[1]

It was clear from the outset—both to Einstein and his contemporaries—that obtaining mathematically exact solutions would be difficult, to say the least. Even if you knew the position, mass, and velocity of every particle within a given space, that still would not allow you to simply determine how such a system would evolve into the future, owing to the nonlinear—and seemingly circular—relationship between matter, energy, and spacetime curvature. You would need to know how matter and energy are both distributed and moving in order to determine how spacetime is curved, but you also need to know how spacetime is curved in order to figure out how matter and energy are distributed and moving. The situation, a hallmark of a nonlinear system, is reminiscent of M. C. Escher's paradoxical lithograph that depicts two hands, each drawing the other into existence—an image that raises profound chicken-versus-the-egg quandaries. A further challenge stems from the fact that in general relativity there is not one equation to be solved but ten interconnected equations that all have to be solved simultaneously. Taking on the equations one at a time would not suffice.

If Einstein's suspicions had been borne out, and no exact solutions were attainable in his theory, one might reasonably question the utility of a set of equations purporting to describe our universe. Had that been the case, it's almost certain that general relativity would not have blossomed into the robust and still-vibrant field it has become over the past 100-plus years. As things turned out, however, Einstein's presentiments were quickly and unexpectedly dispelled. And that was due, in large part, to the fact that his landmark November 25, 1915, paper fell into the right hands—namely those belonging to the astrophysicist Karl Schwarzschild.

Schwarzschild had unorthodox inclinations and a willingness to cast his mind in directions that others had barely considered—characteristics that may have enabled him to achieve the breakthrough soon to be described. "My interest," he wrote, "has never been limited to things situated in space, beyond the moon, but has rather followed those threads woven between them and the darkest zones of the human soul, as it is there that the new light of science must be shone."[2] Those were fitting words, given that a year or so earlier, Schwarzschild had stepped away from his position as director of the Potsdam Astrophysical Observatory and, at the age of forty, volunteered to serve in the German army during World War I. When a copy of Einstein's latest paper was sent to him in December 1915, he was stationed on the eastern front in Russia, where the German army kept him busy calculating artillery trajectories and tackling other ballistics problems. But Schwarzschild had long been interested in the development of general relativity, and somehow, amid his military chores, he found time to read Einstein's paper—perhaps during quiet moments between battles.

Upon gazing at the equations that cast gravity in an entirely new light, Schwarzschild was almost immediately seized with inspiration: What, he wondered, would the gravitational field look like surrounding a point mass—or any compact object—in otherwise empty space? He realized that he might be able to solve the equations of general relativity in this setting, provided that he made a few simplifying assumptions. First, the object in question—which could be a star, for instance, or a planet—had to be spherically symmetric, a supposition that makes the mathematical analysis much less daunting. The spherical object, moreover, had to be nonrotating and otherwise still. The spacetime outside the object, furthermore, would be devoid of matter and energy—a vacuum, in other words. In addition, the spacetime itself would be static, which constitutes another form of symmetry—symmetry under time translation.

In this way, Schwarzschild reduced a problem with four independent variables (t, x, y, z)—what had been a set of so-called *partial* differential equations—to a problem with just one independent variable, r, the radius, leaving him with a set of much-easier-to-solve ordinary differential equations of one variable. This simplification was possible because time (t) was no longer a variable and, owing to spherical symmetry, the only thing that mattered was r, the distance from the center of the star; the directionality—and the x, y, z coordinates of a given point in spacetime—no longer had any bearing on the relevant physics. Schwarzschild still had nonlinear equations to solve, but they turned out to be much easier and, in fact, quite doable—at least for him.

He contacted Einstein in a letter dated December 22, 1915, which described the gravitational field outside of (though not including) a point mass—a result that constituted the first exact solution obtained to the field equations of general relativity, which had been presented just four weeks earlier. It was, to be clear, a mathematical solution to the linked, nonlinear differential equations—albeit one that also greatly illuminated the physics of the situation.

"In order to become versed in your gravitation theory," Schwarzschild explained in his letter to Einstein, "I have been occupying myself more closely with the problem you posed in the paper on Mercury's perihelion," adding that "I took my chances and made an attempt at a complete solution,"[3] and a complete solution is what he got. In his setup, the sun served as the spherical point mass, and his determination of the spacetime geometry around the sun enabled him to work out the precise mechanics of Mercury's orbit.

"It is a wonderful thing that the explanation for the Mercury anomaly emerges so convincingly from such an abstract idea," Schwarzschild told Einstein. "As you see, the war is kindly disposed toward me, allowing me, despite fierce gunfire at a decidedly terrestrial distance, to take this walk into your land of ideas."[4]

It should be noted that Schwarzschild's solution is by no means limited to working out the details of Mercury's previously baffling motions; it's far more general than that. His solutions apply equally well to the orbits of other planets around the sun and indeed to the trajectories of stars and planets around any spherical and strongly gravitating body. At great distances from such a body, gravity behaves just as Newton's laws say it should. But close to a large, massive object, Schwarzschild's equations illustrate the differences in the workings of gravity that arise from general relativistic effects.

Einstein responded to Schwarzschild in a January 9, 1916, letter. "I examined your paper with great interest," Einstein wrote. "I would not have expected that the exact solution to the problem could be formulated so simply. The mathematical treatment appeals to me exceedingly."[5] Einstein was so enthusiastic, in fact, that he submitted Schwarzschild's paper to the Prussian Academy on January 13, and it was published three days later.[6]

In February, Schwarzschild published a follow-up paper in which he provided a mathematical description of the interior of a simple model of a star, in this case, "a homogenous sphere of finite radius, which consists of incompressible fluid. The addition 'of incompressible fluid,' is necessary," he wrote, "because in the theory of relativity, gravitation depends not only on the amount of matter but also on its energy."[7] If the fluid were compressible, it would be capable of storing energy, and thus changing its gravity.

What he found inside his model star, moreover, was truly astounding: If a star's mass, M, were packed into a small enough spherical region (of radius r)—if, in other words, M/r exceeded a threshold value—then nothing, not even light itself, could escape the star's intense gravitational pull. Any matter or energy drawn inside that critical radius, called the Schwarzschild radius, would not be directly observable because any light that it might emit or reflect would instead be stuck inside.

The value of this radius is a direct consequence of the Schwarzschild metric, the formula that Schwarzschild came up with for determining distances in the spacetime surrounding a spherical mass. The metric has a term whose denominator equals $1 - 2m/r$. The Schwarzschild radius, r_s, occurs when $r = 2m$ (where $m = GM/c^2$ and G is the gravitational constant, M the mass of the star, and c, of course, the speed of light). Something very strange occurs when r equals $2m$ because at that point the denominator of one term goes to zero, and the equation becomes ill-behaved. (For the sun, or any spherical star of equal heft, this would happen if all its mass were confined within a radius of about 3 kilometers, instead of its actual value, which is just shy of 700,000 kilometers.) The sphere bounded by the Schwarzschild radius is now referred to as the event horizon, and it can be thought of as the point, or surface, of no return. You can go in easily enough, but—as with the proverbial roach motel—you can't get out. Or, to paraphrase the familiar Las Vegas adage, what happens inside the event horizon stays inside the event horizon.

Furthermore, the Schwarzschild metric also contains the term $2m/r$, and as one approaches the star's centermost point, where $r = 0$, that term goes to infinity—as would the star's density and pressure. Such a point is called a singularity, and it is at places like this that the theory of general relativity breaks down and its predictions become unreliable.

Einstein believed that an object with such bizarre properties—enveloped by a weird event horizon and harboring an even more unfathomable singularity at its core—was a mathematical artifact that could not actually exist in the universe. It might have just been a byproduct of the perfect spherical symmetry assumed in Schwarzschild's derivation—conditions that would not be seen in nature. "If this result were real, it would be a true disaster," Einstein claimed.[8] The astronomer and general relativity advocate Arthur Eddington was another skeptic, who said, "There should be a law of nature to prevent a star from behaving in this absurd way." Any claims to the contrary, Eddington argued, should be regarded as acts of "stellar buffoonery."[9]

Schwarzschild, reportedly, had his own doubts regarding the physical reality of these objects that, fifty years later, came to be known as black holes, as he couldn't think of a viable mechanism for how such things could be formed.[10] But sadly, Schwarzschild barely had any time to delve into this matter further, as he died a few months later, on May 11, 1916, at the age of forty-two.

Nevertheless, Schwarzschild had laid the mathematical groundwork for black holes—about a half century before any credible empirical evidence was available in support of their existence. In the meantime, inquiries into the nature of these hypothetical objects continued to be strictly theoretical and largely mathematical.

In 1923, the mathematician George David Birkhoff extended Schwarzschild's result in a theorem named after him. Birkhoff proved that the gravitational field outside any spherical distribution of matter is *uniquely* described by the Schwarzschild solution to the vacuum Einstein field equations, where $G_{ij} = 0$.

Birkhoff's result was stronger than Schwarzschild's because the earlier solution only pertained to a static spacetime that was unvarying in time, whereas Birkhoff's theorem applied to a so-called time-dependent spacetime that could change. Birkhoff showed, in other words, that "Schwarzschild's solution represents the gravitational field outside any spherically symmetric body, evolving in any manner whatever," explained the mathematician Demetrios Christodoulou. And that was even true for a spherical star that was undergoing gravitational collapse after having exhausted the fuel needed to sustain nuclear fusion.[11]

Although mathematics had raised the possibility of black holes, the concept remained an abstraction until a persuasive argument could be made as to how objects of this sort might actually materialize in the real world. A September 1939 paper by the physicist J. Robert Oppenheimer and his student Hartland Snyder, "On Continued Gravitational Contraction," offered a compelling answer. After studying solutions to the field equations of general relativity, Oppenheimer and Snyder showed how sufficiently massive stars, "which have used up their nuclear sources of energy," could undergo uncontrollable, and catastrophic, gravitational collapse. And if a star were unable to rid itself of mass by means of radiation, "this contraction will continue indefinitely," they wrote. "The star thus tends to close itself off from any communication with a distant observer; only its gravitational field persists."[12]

Even though the model problem Oppenheimer and Snyder chose to focus on was "highly idealized," commented Christodoulou, "their work was very significant, being the first work on relativistic gravitational collapse." And in this way, guided solely by Einstein's equations, they showed how a black hole could be formed, thereby bringing the hypothetical notion, which sprang from Schwarzschild's wartime musings, closer to the realm of plausibility.[13]

Ironically, one month later, in October 1939, Einstein published a paper in the *Annals of Mathematics* that presented an analysis also based on his theory of general relativity, yet he arrived at a very different conclusion. "The essential result of this investigation," Einstein wrote, "is a clear understanding as to why the 'Schwarzschild singularities' do not exist in physical reality."[14]

At that juncture, the astrophysical status of black holes would have to be regarded as debatable. Mathematicians, nevertheless, continued to make progress—one of whom was John L. Synge. In a 1950 paper, Synge "was able, for the first time, to penetrate and explore fully the region inside the so-called Schwarzschild radius, what we now call a black hole," commented the mathematician Petros Florides. "At a time when many relativists, including Einstein, thought that it did not even make sense to talk about this region, this work is very remarkable indeed."[15]

Synge used geometric techniques to demonstrate that the assumed singularity at the Schwarzschild radius (what we now call the event horizon) was not an *actual*, as in physical, singularity—nor was it a place where spacetime comes to an abrupt end—but was merely a *coordinate* singularity, one that artificially cropped up due to the choice of coordinates in the Schwarzschild metric. Synge was the first to show, by geometric means, how the Schwarzschild solution could be maximally extended, which basically meant that any geodesic or path a particle could take, starting from any point in spacetime, would extend smoothly and infinitely in both directions, unless that trajectory ended in an actual singularity that could not be eliminated by a change in coordinates. In this case, the methods he used were at least as important as the results he obtained, given that geometry— in contrast to the more traditional tools of calculus—could afford a global perspective, a way of looking at the whole of spacetime, rather than little pieces of it.

From the outset of his work in general relativity, Synge embraced the geometric approach introduced by Hermann Minkowski. And Synge's impact on the field extended well beyond his extension of the Schwarzschild solution, Florides maintained: "The almost universal geometrical approach to the theory of relativity that began in the 1960s is due primarily to Synge's influence." Synge concurred with that assessment, albeit in his customarily modest way, commenting in 1972 that "if you were to ask me what I have contributed to the theory of relativity, I believe that I could claim to have emphasized its geometric aspect."[16]

One decade earlier, at a landmark 1962 Gravitation and General Relativity meeting in Warsaw, Synge provided a geometric perspective on the movements of objects in the presence of massive bodies with correspondingly strong gravitational fields. Sitting in the audience—among such luminaries as the Nobel Prize–winning physicist Paul Dirac and Richard Feynman, who would soon win a Nobel Prize himself—was Roy Kerr, a twenty-eight-year-old mathematician from New Zealand. Another speaker, Vitaly Ginzburg (who also went on to win a Nobel Prize in Physics), stressed that any strongly gravitating body would necessarily be spinning—in the same way that every star and planet ever observed rotates—so that it was incumbent upon the research community to take up the issue of rotational effects in the general relativity field equations. These words made a powerful impression on Kerr, who returned with a clear sense of purpose to his research institution—the University of Texas, where he had a one-year visiting appointment at the newly formed Center for Relativity.[17]

Kerr wasn't the only one pursuing this problem. The physicist Ezra Newman was also looking into the case of rotating black holes. A big challenge related to rotating objects—black holes, stars, or planets—stems from the fact that they are oblate, meaning that they have bulges at the center and are, therefore, not spherically symmetric. That makes solving the Einstein equations much more complicated. In the course of his calculations, Newman had come to believe that no solutions could be found for rotating black holes. A Texas colleague, Alan Thompson, advised Kerr not to waste his time on this problem since Newman had already proved that no solutions exist. Kerr, however, did not give up so easily. He read through Newman's paper and quickly realized that Newman had made a mistake. That gave him hope, and he poured himself into the problem over the next few weeks, which were described by the cosmologist (and Kerr biographer)

Fulvio Melia as "a furious cocktail of adrenaline, trance-like bouts of distraction, and the smoke from seventy cigarettes a day."[18]

To simplify his task so that it might, in fact, be achievable, Kerr made the following assumptions. Although the black hole in question was rotating—unlike an unmoving Schwarzschild black hole—Kerr posited that this black hole would rotate at a steady, uniform rate, so that nothing would be changing in time. Second, rather than being a perfect sphere, the black hole would instead be fatter at the equator, having the shape of an oblate spheroid. But even though the black hole was not spherically symmetric, he maintained, it would still be axisymmetric—meaning it has symmetry around its axis of rotation. Furthermore, Kerr assumed that the tensor in the Einstein equations representing the curvature of free space devoid of matter—the only portion of Riemannian curvature that persists in a vacuum—would have simple properties that, in turn, make those equations easier to solve. Mathematicians call this last condition, pertaining to space-time curvature, algebraic specialty. After adopting those stipulations, Kerr wrote down a version of Einstein's equations that had two free (adjustable) parameters: the same mass parameter that appeared in the Schwarzschild metric, plus the system's angular momentum or spin. Setting up the problem this way, he was able to find an exact solution, thereby describing the spacetime curvature outside a spinning black hole. The result he arrived at was a generalization of the Schwarzschild solution because when the rotation is set to zero, the Kerr solution reduces to the Schwarzschild case.

Kerr later credited his success on this long-standing problem to "a maverick streak, a lack of attachment to the traditional ways of doing things, and—truth be told—a fairly mischievous skepticism of what I read in books and heard from elders.... Untethered to any established ideas, I felt free to question what I was told, to criticize anything and everything that I believed was wrong, and to pursue paths that others might have shunned. It was precisely this intellectual flexibility that allowed me to discover the mathematical expression for the spacetime surrounding a spinning object."[19]

The result was epic, even though the paper he submitted to *Physical Review Letters* in July 1963 (and published about five weeks later) filled less than one and a half pages of the journal.[20] This was, according to Melia, "a breakthrough solution to Einstein's equations of general relativity that [had] defied the greatest scientific minds of the twentieth century."[21] The Kerr solution was hugely important because the black holes he described—now called Kerr black holes—are the best mathematical representations we have of real black holes and their associated physical properties. Among the properties that Kerr discovered was that a spinning black hole, like the non-spinning Schwarzschild variety, would be enveloped within a surface called an event horizon. And the singularity at the heart of a Kerr black hole would, owing to the object's angular momentum, assume the form of a ring rather than a point. Although an object of this sort might sound freakishly odd, it was proved in the mid-1970s by Brandon Carter and others that the Kerr solution was very broad in scope, as it applies to any kind of spinning black hole that could conceivably exist.[22]

In a lecture given at the University of Chicago in 1975, the Nobel Prize–winning physicist Subrahmanyan Chandrasekhar was effusive in his praise of Kerr's contribution. "In my entire scientific life, extending over forty-five years, the most shattering experience has been the realization that an exact solution of Einstein's equations of general relativity, discovered by the New Zealand mathematician Roy Kerr, provides the absolutely exact representation of untold numbers of massive black holes that populate the universe," Chandrasekhar said. "This incredible fact that a discovery motivated by a search after the beautiful in mathematics should find its exact replica in Nature, persuades me to say that beauty is that to which the human mind responds at its deepest and most profound."[23]

In the fall of 1963, Roger Penrose—a mathematician-turned-physicist who was drawn from algebraic geometry to general relativity after reading Synge's book on the subject[24]—came as a one-year visiting professor to the University of Texas, where he engaged in many conversations with Kerr. A year later, after taking a position as reader of applied mathematics at Birbeck College in London, Penrose began to wonder whether singularities were intrinsic, unavoidable properties of Schwarzschild and Kerr spacetimes.[25] He also wondered whether singularities could arise in objects that lacked those same symmetries.

The more generalized, asymmetrical situation was much harder to take on because of the previously cited difficulties involved in solving the Einstein field equations. But rather than trying to solve those equations outright for different special cases, Penrose developed a new set of mathematical tools—from geometry and topology—to analyze the properties of spacetime. The results of his inquiry were presented in a three-page January 1965 paper, "Gravitational Collapse and Space-Time Singularities," which, according to the physicist

Werner Israel, "has claims to be considered the most influential de-velopment in general relativity in the 50 years since Einstein founded the theory."[26] The importance of this paper not only stemmed from the specific result that Penrose had obtained but was also due to the mathematical approach he pioneered, which has led to a new era of study in general relativity.

In his "Gravitational Collapse" paper, Penrose asked whether the Schwarzschild singularity "is, in fact, simply a property of the high symmetry assumed." The same might be said for the Kerr solution, he added, "since a high degree of symmetry is still present (and the solu-tion is algebraically special), it might again be argued that this is not representative of the general situation." His analysis addressed, head-on, the issue of "collapse without assumptions of symmetry," and he ended up proving that "deviations from spherical symmetry cannot prevent space-time singularities from arising."[27]

What Penrose proved, to be precise, was that the event horizon of either a Schwarzschild or Kerr black hole is something he called a closed trapped surface. He showed, moreover, that when a closed trapped surface is formed, the collapse to a singularity is the inevi-table result, regardless of symmetry considerations.

This latter point is especially important because, up until that time, many people in the field of general relativity—including the noted doubter, Einstein himself—thought that a singularity could only exist, if it could exist at all, because of a high, and perhaps un-naturally high, degree of symmetry. Penrose demonstrated that was not the case: that the emergence of black holes was not contingent upon unrealistic symmetry assumptions. In so doing, he went a long way toward convincing other scientists, many of whom were previously skeptical, that the formation of such objects could be a genuine physical phenomenon.

So what exactly is a closed trapped surface—the concept Penrose introduced, drawing on arguments rooted in geometry and topology? It's basically a surface whose curvature is so extreme that light gets trapped inside. Light rays are unable to escape, or even move in the outward direction, and instead get forced inward. In particular, Penrose looked at what would happen to those rays in the vicinity of a black hole. Imagine that light is emitted from a spherical surface that surrounds, and is just outside of, a black hole. That light can travel in two directions: It could go away from the black hole, or it could go inside. But now imagine that this sphere is inside a black hole or a star that was undergoing uncontrollable gravitational collapse. Either way, the area of that surface will steadily and inexorably decrease. And instead of fanning out in all directions, light rays emanating from that surface will be bent back and forced to converge toward the center.

"Since the surface of the region shrinks to zero, so too must its volume," Stephen Hawking, Penrose's colleague, wrote. "All the matter in the star will be compressed into a region of zero volume, so the density of matter and the curvature of spacetime become infinite."[28] To inject a note of caution here, it should be said that the notion of curvature going to infinity at the singularity is—at the moment—a widely held belief that no one, as of yet, has been able to prove, nor have there been any credible attempts to do so. If this tenet is indeed correct, then at the place where the curvature becomes infinite, an erstwhile moving particle or light ray can proceed no further, and it

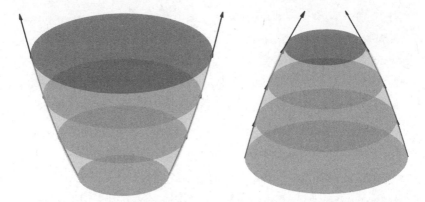

A trapped surface (*right*) is somewhat like an inverted light cone

might be said that spacetime itself comes to an end. It is at just such a spot, where a "tear" suddenly appears in the fabric of spacetime, that a singularity is destined to form.

Penrose demonstrated, moreover, that a closed trapped surface is stable in the sense that it can withstand a small disturbance or perturbation. Establishing stability from a mathematical standpoint is a critical, make-or-break threshold: Although objects like closed trapped surfaces that give rise to singularities could exist on paper—mathematically, that is—if they were unstable, they would have no physical significance and would never be seen in the real world. The caveat here is that Penrose proved the stability of the closed trapped surface but said nothing about the structure of the singularity itself. He showed that the existence of some singularity would survive a perturbation and that a black hole would be the invariable result. But the structure of the singularity itself could potentially change and emerge from the transaction with different properties than those originally in place.

The reason closed trapped surfaces are inexorably fated to become black holes has to do with positive curvature, which is something that such surfaces have in the extreme. Being inside a closed trapped surface is like being in a room in which the roof, walls, and floor are closing in from all directions—not a great place for someone who is claustrophobic! Light rays, even those initially heading outward, get turned around by this intense curvature and redirected inward.

"If the surface area is initially decreasing, it will continue to decrease because there's a focusing effect," the mathematician Richard Schoen explained. "You can also think of great circles on the globe that start at the north pole and separate, but because the curvature is positive on a sphere, the lines start to converge and eventually come together at the south pole. Positive curvature gives you this [same] focusing effect."[29] And the positive curvature of a closed trapped surface provides the ultimate in focusing.

"Penrose's singularity theorem acquired its amazing power from a new mathematical tool that he used in its proof," wrote Kip Thorne, "a tool that no physicist had ever before used... in general relativistic calculations: topology"[30]—a branch of mathematics that deals with the general shape of objects as opposed to their exact shape or geometry. In addition to the "singular" import of his theorem, Penrose's introduction of an entirely new, topological approach helped transform the study of general relativity.

Around that time, Penrose gave a talk about his theorem at Princeton where the physicist Robert Dicke told him: "You've done it, you've shown general relativity is wrong!" Penrose insisted that was absolutely not the case. "What I proved... doesn't mean general relativity is wrong," he said. "But you do have to have singularities."[31] This was a point that some physicists, including Einstein, had been hesitant to accept.

Not to take anything away from Penrose's far-reaching contribution(s), it should be pointed out that his 1965 theorem did not, in itself, prove the existence of black holes. He proved that closed trapped surfaces, once formed, will degenerate into objects from which light cannot escape—objects containing a central singularity that were just then starting to be called black holes.[32] Although that was a momentous feat—which marked an original and radical break in general relativity research—the theorem does not spell out exactly what it takes to make a trapped surface in the first place. A 1979 theorem—which was published in 1983 by Schoen and Shing-Tung Yau[33]—carried things a step further.

Physicists had long assumed that a black hole would form once the matter density in a particular region reached a high enough value, but that conviction was based on rather vague arguments that had never been quantified or formulated in a definitive way. Schoen and Yau set out to determine the precise conditions that would give rise to a trapped surface. Their theorem showed that when the matter density of a given region is twice that of a neutron star (a star in which electrons and protons have been forced together by gravity, which is, itself, roughly 100 trillion times denser than water), a trapped surface will form and the object will directly collapse to a black hole rather than to some other state.

The work of Schoen and Yau, commonly referred to as the black hole existence proof, was mathematically rigorous, just as Penrose's singularity theorem was mathematically rigorous. But whereas

Penrose's arguments were largely topological (having to do, in some cases, with differences between flat Euclidean space and the surface of a sphere), Schoen and Yau relied on geometric arguments involving the mean (or extrinsic) curvature of so-called minimal surfaces—surfaces that have the smallest possible surface area for a given boundary.

This exemplified just one of many lines of mathematical inquiry related to black holes that emerged in the wake of Penrose's singularity theorem. In a way, it is rather staggering to consider the number of ideas that have been sparked, and the depths to which they have been pursued, since the general notion of what we now call black holes first appeared, unexpectedly, in Schwarzschild's 1916 solution to the Einstein equations. Among the active research directions today are such questions as: What shape or shapes can a black hole assume? Can black holes exist in a spacetime of more than four dimensions, and if so, what properties would they have? Must black holes always be surrounded by event horizons, or can uncovered, "naked" singularities exist? Are Kerr black holes truly stable—entities that can withstand some degree of disturbance and return (almost exactly) to their original state? And are all Kerr black holes of a given mass, spin (or angular momentum), and electric charge essentially the same and hence indistinguishable from one another—regardless of how the black holes were made or what materials (and recipes) went into making them?

A 1972 paper by Hawking, which addressed the first question above, introduced a topological theorem that affirmed that the surface of a black hole must be a sphere.[34] In order to prove this, Hawking relied on the Gauss–Bonnet formula from the 1800s, which relates the curvature of a surface to its underlying topology. In slightly more technical terms,

Hawking's theorem says that the surface of a four-dimensional black hole—which is typically referred to as the event horizon—is a three-dimensional object, and a cross-section, or slice, of that surface must be a two-dimensional sphere. (And spheres with two-dimensional surfaces are the kind of spheres we're familiar with in everyday life.)

From a technical standpoint, however, it's more accurate to call the surface described in this theorem the *apparent* horizon rather than the event horizon. The apparent horizon is the outermost trapped surface surrounding a black hole. It is the same as the event horizon in the case of Schwarzschild or Kerr black holes but, as a general matter, the two surfaces do not have to coincide.

Four-dimensional black holes—those with three spatial dimensions and one of time—"have a number of remarkable quantities," the physicist Gary Horowitz wrote in 2005. "It is natural to ask whether these properties are general features of black holes or whether they crucially depend on the world being four-dimensional." In fact, Horowitz added, "many of them are indeed special properties of four dimensions and do not hold in general."[35] He cited a 2002 paper by the physicists Roberto Emparan and Harvey Reall, who found solutions to the Einstein equations that offered the first examples of five-dimensional black holes. Emparan and Reall called these rotating toroidal (or donut-shaped) objects black rings, demonstrating that a black hole in five-dimensional space was not subject to the restrictions imposed by Hawking's theorem and that its event horizon could therefore have non-spherical topology.[36]

A 2006 paper by Schoen and his mathematics colleague Gregory Galloway generalized Hawking's 1972 theorem, showing again that, in higher dimensions, a black hole does not have to be a sphere. Other topologies are indeed possible, but these objects must have a special kind of positive curvature (based on an argument made in the Schoen–Yau proof of the positive mass conjecture, which will be discussed in Chapter 7) that forces them to curl up rather than spread out. This "special" curvature is, in fact, positive scalar curvature—where scalar curvature, itself, is a generalization of two-dimensional intrinsic curvature, as originally defined by Gauss, to a space or manifold of any dimension.

Galloway and Schoen demonstrated that there are just three allowable shapes for the event horizons of five-dimensional black holes: a three-dimensional sphere, a so-called ring (a shape, first discussed by Emparan and Reall, made by mathematically combining a circle and a two-dimensional sphere), or a class of objects called lens spaces (which are made by folding up three-dimensional spheres in complicated ways).[37] Since that time, mathematicians have begun to explore even higher-dimensional black holes—work that continues to this day.

In the meantime, physicists have investigated the possibility of whether exotic, higher-dimensional black holes might be found in nature. Moreover, they have identified potential signatures that tiny versions of these objects would give off during their brief lifetimes—a tiny fraction of a second, in all—if they were produced, fleetingly, at the Large Hadron Collider or other particle accelerators. The detection of so-called quantum black holes in high-energy physics experiments would support an idea proposed by Hawking in 1971. And if any of these objects were not spherical in shape—but instead had different, unusual topologies—that would provide tantalizing indications of higher dimensions because four-dimensional black holes, in accordance with Hawking's theorem, must be spheres.

In 1969, Penrose launched another avenue of investigation, which has long kept researchers occupied, by advancing what's called the weak cosmic censorship conjecture.[38] It states that all singularities formed from the gravitational collapse of matter are hidden inside black holes and barred from external view by the event horizons that envelop them. Hawking described the idea in somewhat more colorful terms: "God abhors a naked singularity," he said, noting that "at the singularity, the laws of science and our ability to predict the future would break down."[39]

A decade later, Penrose issued the strong cosmic censorship hypothesis, which maintains that general relativity is a deterministic theory, meaning that, starting from initial conditions, predictions can confidently be made into the future. Despite the name, the "strong" cosmic censorship conjecture is not literally stronger than the "weak" conjecture. It is, however, a much broader statement. And the two conjectures are different in the sense that there are examples for which one case may be upheld and the other violated.

With regard to the strong hypothesis, Penrose aimed to bypass the problem of general relativity's potential breakdown inside a black hole by postulating that spacetime comes to an abrupt end at what's called the Cauchy horizon—a hypothetical boundary that lies inside the event horizon. In this way, general relativity can successfully carry us to the edge of spacetime and no farther. But if the theory is not intended to go beyond that point, as Penrose surmised, general relativity's predictions would indeed be trustworthy within the domain it is supposed to describe.

Both versions of Penrose's censorship conjecture are intended to preserve the reliability of general relativity—and thereby safeguard scientific determinism—by cordoning off singularities into a kind of lockbox that would enable calculations about spacetime, as one moves into the future, to be confidently carried out. In this way, the predictive power of physics, and that of general relativity in particular, could be saved from the chaotic effects of uncloaked singularities. "Penrose came up with a conjecture that basically tried to wish this bad behavior away," said the mathematician Mihalis Dafermos.[40]

Penrose's censorship conjectures were not precisely stated, at least not from a mathematical standpoint, which has led to critical reappraisals of these statements. Researchers have subsequently shown that certain versions of the conjectures do not hold up. In the 1980s, for example, Christodoulou found circumstances in which naked singularities could exist—counterexamples to some special cases of weak cosmic censorship—but the general case is still unresolved, remaining one of the biggest open questions in general relativity.

In 2017, Dafermos and his colleague, the mathematician Jonathan Luk, disproved a form of the strong censorship conjecture, using the example of a Kerr (uncharged, rotating) black hole. They took aim at a critical contention in Penrose's argument—namely that the Cauchy horizon within the black hole's interior is a singularity in itself, marking the end of spacetime. Dafermos and Luk proved that the Cauchy horizon is actually a "weak" singularity, which is to say that it does not mark the end of spacetime; a particle could, in fact, sail right past it. By contradicting a key assertion made by Penrose, they showed that strong cosmic censorship does not prevail under the conditions they investigated,[41] which relate to a very strong statement of the conjecture. Nevertheless, weaker versions of the (strong) conjecture have not been ruled out and still warrant investigation.

Meanwhile, there has been an even longer-standing mystery, dating back to Kerr's 1963 proof that revealed a solution to the Einstein equations in the form of rotating black holes, subsequently dubbed Kerr black holes. The question is, are these objects stable? What stability means here, explained Sorbonne University mathematician Jérémie Szeftel, "is that if I start with something that looks like a Kerr black hole and give it a little bump"—by throwing some gravitational waves at it, for instance—"what you expect, far into the future, is that everything will settle down, and it will once again look exactly like a Kerr solution."[42] The problem of stability, added the mathematician Sergiu Klainerman, "is not only a deep mathematical question but one with serious astrophysical implications. Indeed, if the Kerr family would be unstable, black holes would be nothing more than mathematical artifacts—mathematical ghosts, as I like to say."[43] Discovering "some mathematical instability when perturbing black holes," added the physicist Thibault Damour, "would have posed a deep conundrum to theoretical physicists and would have suggested the need to modify, at some fundamental level, Einstein's theory of gravitation."[44]

One reason the question of stability has remained open for so long is that most explicit solutions to Einstein's equations, such as the one found by Kerr, are stationary, referring to black holes that can rotate at a steady rate but don't otherwise change in time. But the black holes we see in nature are dynamic objects that can accrete matter, spew out radiation of all types, and emit powerful jets of white-hot plasma. To assess stability, researchers need to subject black holes to minor disturbances and then find out what happens to the solutions that describe these objects as time moves forward.

A stable system is one in which small changes have small consequences. For example, suppose you have two seemingly identical model rockets. First, you launch one from a certain spot and at a precise angle. Next, you shift the launch spot for the second rocket by an inch and change the ultimate velocity by altering the launch angle by one tenth of a degree. If the two rockets end up following very similar trajectories, you'd then conclude that the system was stable for small perturbations. In general relativity, the change in launch spot would take place in four dimensions rather than just one, and the momentum tensor would have to be adjusted instead of just modifying the velocity.

To take another example, imagine sound waves bouncing off a wineglass. Typically, the waves will vibrate the glass a little bit, and the glass would then settle down. But if someone were to sing loudly enough and at a pitch that exactly matches the glass's resonant frequency, the glass could shatter. It clearly would not go back to its initial state—that of a normal, intact wineglass. This scenario would instead depict an instability, an instance of a small disturbance leading to a very different endpoint.

Klainerman and Szeftel joined forces with Elena Giorgi in a 2022 paper to see whether a similar resonance phenomenon could happen when a black hole is struck by a wavelike disturbance. The trio was able to complete work carried out by Klainerman and Szeftel in several prior papers, thereby proving that slowly rotating Kerr black holes are stable.[45]

The three mathematicians relied on a strategy called proof by contradiction, employing an argument that went roughly like this: The conjecture they set out to prove holds that the Kerr solution exists for all time, even when subjected to slight perturbations. So they assumed that, to the contrary, "the solution does not exist for all time—that, instead, there is a maximal time, T_{max}, after which the solution does not exist because the perturbation becomes too strong," Giorgi explained. "We then use some mathematical trickery"—an analysis of partial differential equations—"to show that we can extend the time beyond T_{max}."[46] Their initial assumption was thus contradicted, implying that the conjecture itself must be true.

On the basis of this work, stability was only proven for slowly rotating black holes—those for which the ratio of angular momentum, a, to the mass, m, is much less than one. Exactly how much less, they could not say. Alternatively, when a/m is equal to one, a black hole would be classified as extremal, meaning that it is spinning at the maximum theoretically allowable rate. The 2022 paper pertained to a black hole whose angular momentum is somewhere at the other end of the spectrum. A "full resolution" of the Kerr stability conjecture, which would apply to any physically possible value of angular momentum, has not yet been attained.

Meanwhile, there are other questions concerning Kerr black holes that remain unresolved. One, which is sometimes called the uniqueness (or rigidity) conjecture, maintains that the only solution to the Einstein equations that yields a stationary or uniformly rotating black hole is, in fact, the Kerr solution. Beyond that lies an even broader, and much more daunting, proposition known as the final state conjecture, which basically says that if we wait long enough, all matter will be contained within Kerr black holes. And ultimately, the conjecture holds, the universe will evolve to a finite number of Kerr black holes moving away from each other, and nothing else will be left except for some gravitational radiation.

Given that the final state conjecture depends on Kerr stability and uniqueness, as well as on weak cosmic censorship and other conditions—and since no one yet knows how to tackle this problem, or even approach it in its full sweep—we'll focus, for the moment, on the uniqueness conjecture, which is also referred to as the no-hair theorem (though this "theorem" should technically be called a conjecture, as it has been only partially proven, at best).

The term *no-hair* might sound confusing. What it means is that an isolated Kerr black hole is completely characterized by just two numbers: its mass and spin. Beyond that, there are no additional defining features—no "hair," so to speak—that would enable an observer, looking from afar, to differentiate between two black holes of equal mass and spin. This would be true, the theorem says, even if the objects were constructed out of very different kinds and combinations of material—elementary particles, dust, stars, and the like—that carried with them very different histories. That raises the question of whether information might be lost, or destroyed, when material is drawn into a black hole. According to quantum theory, information should be conserved, which is why this scenario is referred to as the black-hole information paradox.

Conversely, if two black holes of equal mass and spin can be distinguished from one another, it means their surfaces (or event horizons) are not entirely featureless; some information or "hair" must be retained to enable an outside observer to tell them apart. If information is saved in this fashion, the so-called paradox associated with it would then go away.

Not to complicate our discussion too much, but the no-hair theorem is commonly ascribed not just to a Kerr black hole, but also to a Kerr–Newman black hole, which is basically a Kerr black hole (of a certain mass and spin) that has a designated electric charge as well. The Kerr–Newman solution describes the spacetime geometry around a charged, rotating mass. And in this milieu, it is said that a black hole is completely, and uniquely, characterized by *three* observable properties: its mass, spin, and charge.

The first proofs of the no-hair theorem were offered in the early 1970s by Brandon Carter, David C. Robinson, and Stephen Hawking. Their arguments came in two parts. Carter and Robinson first showed that for a steadily rotating, axisymmetric black hole, the Kerr solution is unique. Hawking proved separately that if the black hole's event horizon is smooth in a special way—characterized by a property known as real analytic—then the black hole must be axisymmetric.

But these proofs had some nontrivial limitations. Many researchers in the field consider Hawking's assumption regarding horizon smoothness to be very strong and perhaps overly strong: The real analytic condition basically says that the local geometry of an object—the geometry that pertains to just a small region of a surface or manifold—actually controls the global, or overall, geometry. Such an assertion is hard to justify. Moreover, both proofs assumed a vacuum setting in which a black hole could not accrete mass in any way. All indications suggest that there often is matter in the vicinity of black holes—a consequence of their immense gravitational drawing power—meaning that even if the proof of the no-hair theorem in a vacuum was airtight, the same would not necessarily hold for real astrophysical black holes that are surrounded by matter.

In 1988, a counterexample was found by the mathematician Robert Bartnik and his student John McKinnon.[47] Bartnik and McKinnon first coupled the Einstein equations of gravity to the nonlinear Yang–Mills field equations—equations that govern the strong and weak nuclear forces, as well as the behavior of their associated particles. From those coupled equations, a different kind of black hole would arise, a so-called Bartnik black hole, that does not comply with the no-hair theorem. They did this by adding matter to the right-hand side of the Einstein equation, and that matter—which assumes the form of the Yang–Mills field and its accompanying particles—is incorporated into the stress-energy or energy-momentum tensor, T_{ij}. (In a vacuum, as you may recall, T_{ij} equals zero.) In the presence of a Yang–Mills field, explained the mathematician Felix Finster, "a black hole is not characterized only by its mass and spin. It has something extra, particles or fields." It has "hair," in other words.[48]

Bartnik and McKinnon's result pertains to a static (non-rotating) black hole that is spherically symmetric—somewhat analogous to a Schwarzschild black hole. The no-hair theorem, as stated above, pertains to Kerr or Kerr–Newman black holes. But if the theorem is true, it also must apply to Schwarzschild black holes, which are really just special cases of the Kerr solution, involving spherical symmetry and zero spin, as well as zero charge. But unlike Schwarzschild black holes, the black holes that Bartnik and McKinnon conjured up in their calculations had hair, so-called Yang–Mills hair, which meant that two Bartnik black holes of identical mass would look different. They might, for instance, have slightly different clouds of particles hovering around, or clinging to, their respective event horizons, thereby distorting the horizons in a theoretically perceptible way.

One shortcoming of the Bartnik and McKinnon result—at least in the eyes of some mathematicians—is that their counterexample did not come from an actual mathematical proof but rather from a numerical (i.e., computer-based) argument. Starting in the 1990s, Felix Finster, Joel Smoller, Shing-Tung Yau, and other collaborators made the 1988 (Bartnik–McKinnon) work more mathematically rigorous. They proved the existence of static, spherically symmetric black holes with hair, finding an infinite number of such objects—all variations on the single example that Bartnik and McKinnon had presented.

One way to think about this extension of the Bartnik–McKinnon result is to consider our discussion of the action principle in Chapter 3. The Einstein equations can be derived from an action—let's call it A_1—which Hilbert showed in 1915 to be an integral of the scalar curvature. The Yang–Mills equations, which give rise to the Yang–Mills (matter and energy) field, can be derived from a different action, A_2, another integral related to curvature. When these equations are coupled, the actions are actually added to each other: $A_1 + cA_2$, where c is a constant: not the speed of light, but rather a coupling constant that can assume only discrete values. By changing the value of c, you can get an infinite number of solutions, each describing a different black hole. The larger the value of c, the stronger the Yang–Mills field. The key point here is that when these black holes are perturbed by Yang–Mills matter, they will have hair—contrary to the dictates of the theorem that Hawking and company proved in a vacuum setting.

There was a catch, and a sizeable one at that: Mathematicians believed that the black holes derived from the Bartnik–McKinnon solution were unstable and hence could not explain physical phenomena. That was true under the conditions, including that of an extremely smooth event horizon, that were assumed by Hawking. But in 2022, roughly fifty years later, Yau, Yuewen Chen, and Jie Du showed that if that smoothness condition were relaxed, you could get a different class of solutions: hairy black holes that—unlike the black holes derived by Bartnik–McKinnon and Smoller et al.—were actually stable. That, in turn, would imply that the no-hair theorem is not true when matter is present—or at least when matter of the Yang–Mills variety is present.[49]

The work of Yau, Chen, and Du also raised the possibility of a different "final state"—a universe that might ultimately be populated by these newly minted black holes (solutions to the so-called Einstein–Yang–Mills equations) rather than by black holes of the more familiar Kerr variety, as originally theorized.

With this recent addition to the black hole menagerie, there would no longer be just Kerr black holes and Schwarzschild black holes (a special case of the former). If stable black holes could have hair, there would instead be an infinite number of possibilities to choose from—each black hole having a somewhat different hairdo.

Stable black holes of this variety, Chen, Du, and Yau suggested, might have formed in the early universe, surviving to this day as a kind of dark matter—consistent with the notion of the primordial black holes that Hawking started speculating about in the early 1970s.

But these counterexamples to the no-hair theorem, which come in an infinite variety, are still not definitive, given that Chen, Du, and Yau offered numerical solutions, worked out on a computer, that fall short of an actual proof. Chen, Yau, and two other mathematicians are now crafting a mathematical argument that could prove the existence of a family of stable black holes, none of which look exactly alike. However, given that this is still a work in progress, the status of the no-hair theorem has to be regarded as up in the air. That uncertainty, combined with separate quandaries related to spinning black holes, illustrates that Roy Kerr's amazing invention still holds many mysteries.

At least one long-standing question appears to have been cleared up since Kerr's paper came out in 1963: whether black holes exist at all. Despite all the mathematical interest that his result generated, the physical existence of these objects was then very much in doubt. In that year, astronomical evidence and arguments in support of the existence of black holes began to trickle in, although it wasn't necessarily recognized as such straightaway. In 1963, for example, the astronomer Maarten Schmidt discovered quasars, the most luminous objects in the universe, which can outshine the total energy output of our galaxy by a factor of hundreds or even thousands. Six years later, the astrophysicist Donald Lynden-Bell proposed that the source for this monstrous outpouring of energy was gases that had been heated to millions of degrees as they were drawn into a giant—so-called supermassive—black hole that resides in a galaxy core. In 1973, astronomers suggested that a bright source of X-rays called Cygnus X-1 could be a stellar-sized black hole. In 1978, astronomers proposed that the galaxy Messier 87 (M87) harbors a supermassive black hole with a mass of billions of suns.

Even with this mounting evidence, and the plausible interpretations that were attached to it, considerable skepticism about the existence of black holes among astronomers extended well into the 1980s, at least. Finally, in 2019, a global network of radio telescopes called the Event Horizon Telescope captured the first-ever image of the actual silhouette just outside the supermassive black hole residing inside M87. For many, seeing was believing, and the physical reality of black holes was—for them and for most of the scientific community—no longer up for debate.

In 2020, Roger Penrose won a Nobel Prize in Physics for using "ingenious mathematical methods in his proof that black holes are a direct consequence of Albert Einstein's general theory of relativity." His groundbreaking 1965 paper, according to the Nobel Committee, "is still regarded as the most important contribution to the general theory of relativity since Einstein."[50] (Penrose later pointed out that the Nobel citation was "a bit misleading. [It] said that I showed black holes are a robust prediction of Einstein's general theory of relativity. What I really showed is that *singularities* [emphasis added] are a robust prediction of general relativity."[51] That clarification, however, should take nothing away from his prodigious achievement.)

The physicist Sabine Hossenfelder noted that prior to the work of Penrose, and the subsequent singularity theorems he developed with Stephen Hawking, "most physicists thought that black holes were merely mathematical curiosities which appear in general relativity but that they would not exist in reality.... The story of the discovery of black holes demonstrates vividly how powerful pure mathematics can be in the quest to understand nature."[52]

That statement echoed an earlier, more general one made by Hawking himself, who said that "the theory of black holes was developed before there was any indication from observations that they actually existed. I do not know any other example in science where such a great extrapolation was successfully made solely on the basis of thought."[53]

And much of that "thought," as we've seen, was firmly rooted in mathematics. History has indeed shown, on countless occasions, that the power of logic and reasoning can shine a light on places that our observational capabilities have not yet managed to penetrate. In this case, illumination of that very sort came about when people tried to solve an equation—one that its author regarded as possibly unsolvable.

= 5 =

Chasing the Wave

Running in parallel to the development of ideas about black holes was work on a second question that also seemed to spring quite naturally from general relativity. This is not entirely coincidental, of course, as Albert Einstein's paper of November 25, 1915, and the set of equations introduced therein offered an entirely new picture of gravitation and, with it, new phenomena became conceivable that had not been seriously considered before. Black holes were not the only novel possibilities to emerge.

In June 1916, Einstein initiated a second major line of investigation, arguing that just as the presence of matter could cause spacetime to curve, the acceleration of matter could cause spacetime to undulate, creating gravitational waves—ripples, like those produced by a powerboat speeding across a once-still lake, except that they would propagate throughout spacetime at the speed of light.[1]

This was more than just idle speculation on Einstein's part. He was the first person to provide an actual theory of gravitational waves, as well as the first to cast qualitative notions about these waves into explicit mathematical form.[2] And as with the case of black holes, it took about a century to obtain definitive proof of the existence of the postulated phenomenon.

Einstein had doubts as to the physical reality of gravitational waves, just as he had with black holes. He first believed that such waves could be generated by the acceleration of masses in the same way that electromagnetic waves are generated by the acceleration of electric charges. He then decided that would not work because, whereas both positive and negative electric charges existed, there was no such thing as a negative mass. He informed Karl Schwarzschild of his conclusion in a February 19, 1916, letter, which stated "there are no gravitational waves analogous to light waves."[3] But a few months later, in a June 22, 1916, paper, Einstein did predict the existence of these waves, albeit generated by a somewhat different mechanism.[4]

His second paper on the subject, which came out in 1918, corrected an error that had appeared in his 1916 version and served as a foundation for this emerging field.[5] Based on his revised calculations, Einstein determined that gravitational waves would be extremely weak—and too weak, in his estimation, to be detected with foreseeable technology. A 1936 paper he wrote with his assistant, Nathan Rosen, found that any solution to general relativity's field equations that produced gravitational waves would invariably contain a spacetime singularity. In the draft first submitted to *Physical Review*, they concluded that if embracing the concept of gravitational waves meant accepting the physical actuality of singularities, then gravitational waves could not exist after all (although Einstein later made fundamental changes to that manuscript).

In a letter to the physicist Max Born sent that year, Einstein wrote: "Together with a young collaborator [Rosen], I arrived at the interesting result that gravitational waves do not exist."[6] He was even scheduled to give a lecture that year in Princeton on the "Nonexistence of Gravitational Waves," but he softened his stance after a colleague apprised him of an error in his paper with Rosen. During the talk, Einstein said, "If you ask me whether there are gravitational waves or not, I must answer that I don't know. But it is a highly interesting problem."[7]

History has shown that he was wise to take a more moderate position in that talk, and he was also correct in his assessment that gravitational waves would be very faint—and thus very hard to observe. The only hope for detecting these waves, physicists later realized, would be if they were produced in a truly violent, cataclysmic event, such as the high-speed collision and subsequent merger of two black holes. And *violent* is the appropriate word to describe such an occurrence. "Imagine taking 30 suns and packing them into a region the size of Hawaii," said Vijay Varma, a physicist at the Albert Einstein Institute in Potsdam, Germany. "Then take two of such objects and accelerate them to half the speed of light and make them collide."[8] That is a recipe for a clash of epic proportions—and one that could yield detectable traces.

This was not initially obvious. The Schwarzschild solution describes a much calmer situation—that of a stationary, solitary round mass sitting in otherwise empty space with nothing changing in time. The Kerr solution describes another comparatively simple situation—that of a single, not perfectly round mass rotating at a uniform rate in empty space. Again, nothing changes in time. But two rotating Kerr black holes that are sufficiently close together will start circling each other at steadily shrinking orbits and ever-increasing speeds until they slam into each other and ultimately coalesce—continuing to quake long afterward like a bell that's been forcefully struck. Gravitational waves are emitted at every stage of this process, reaching a peak at the moment of the great crash.

The double–Kerr black hole scenario is far more complicated than that of an isolated, motionless black hole. To make the situation somewhat manageable, researchers assume that the local spacetime contains just these two black holes and nothing else—as viewed by a distant observer, located far from the mayhem, in a region where things are tranquil and spacetime curvature is essentially nil. Even with that simplification, a major challenge remains: determining how, in general relativity, gravitational systems evolve into the future.

A key step toward that goal is to formulate and then solve what is known as the initial-value problem, also called the Cauchy problem, for the Einstein field equations. What that means, specifically, is that you start out at an initial time with a particular spacetime geometry—one that satisfies the Einstein equations—and see whether, as the system evolves, you can arrive later at a solution to those same equations. Put more simply, does a solution to the Einstein equations exist as you move into the future—and, if so, how far can you carry those calculations forward?

The question can be expressed in even broader terms: Are the equations of general relativity—which have been solved in just a limited number of special cases, as Schwarzschild did in 1916, and others have since—well-posed? A well-posed question is a mathematically sensible one. The notion of well-posed (and ill-posed) equations was introduced in a 1902 paper by the French mathematician Jacques Hadamard, who laid out three principal criteria:[9] First, a solution to the equation (or equations) has to exist. Second, the solution has to be unique—and by that he meant unique "locally," which is to say persisting for a short period of time. The third criterion relates to the predictability of the equations: If we know the initial condition of a system at some time—such as an object's position, velocity, and angular momentum—can the theory tell us what it will be at a later time? And will small changes in those conditions lead to small changes in the solution itself? Satisfying these latter points is absolutely crucial because if the outcome does not depend on how and where things start off, then causality—the relation between cause and effect that serves as a cornerstone of science—would not prevail.

To sum up, well-posed equations should yield a solution that is unique and not overly sensitive to slight variations in conditions. A system that meets the above specifications, in which things vary continuously—a small change here resulting in a small change there—is considered stable. But it was not known until the middle of the twentieth century, nearly four decades after Einstein formulated his field equations, whether those equations actually fulfilled the standard of well-posedness and whether they could be counted on to yield meaningful results—especially in the extreme, tumultuous circumstances of, say, two black holes hurtling toward each other on a deadly collision course.

In 1952, the French mathematician Yvonne Choquet-Bruhat offered the first proof that the nonlinear version of the Einstein equations would indeed give rise to gravitational waves that travel at a finite speed. Einstein, in his 1916 paper, found wave solutions to a simplified, linear form of the equations based on the assumption that, far from the source of the gravitational field, gravity would be very weak and nonlinear effects could essentially be ignored. "However, he [Einstein] knew that linearization can introduce artifacts that are not present in the general case," explained the mathematician Lydia Bieri.[10] "He knew that when you linearize things, you change things, and you might introduce something that's not there and miss something that is there."[11] Choquet-Bruhat was also the first to prove the well-posedness of the Einstein equations. Her PhD thesis—which was published in 1952 and contained various proofs[12]—has become "one of the most important results in the history of general relativity," according to Bieri.[13] "Much of what we do now [in mathematical relativity] goes back to that 1952 paper."[14]

One of the things Choquet-Bruhat proved was that the Einstein equations are hyperbolic—a kind of partial differential equation (PDE) that behaves like a wave equation. Equations of this sort can, for example, model how waves travel back and forth along a plucked guitar string, a situation involving continual motion and continual changes in time. (There are two other types of PDEs: elliptic equations, which describe space at a single moment in time, showing for instance the gravitational field produced by a given configuration of masses; and parabolic equations, which can describe, among other things, the process of diffusion—such as how a glob of cream gradually distributes itself more uniformly throughout a cup of coffee.)

Choquet-Bruhat built on a brand-new result of the mathematician Jean Leray, her postdoctoral supervisor at the Institute for Advanced Study, who proved in 1952 that a particular class of hyperbolic PDEs are well-posed. Choquet-Bruhat's strategy was to show that the Einstein equations fall into this same category of hyperbolic equation. While there are ways of writing down the Einstein equations that are not hyperbolic, she figured out how to write them in the appropriate hyperbolic form, and that, in turn, enabled her to draw upon Leray's result.

The challenge that Choquet-Bruhat faced involved picking the right coordinate system that would give the field equations especially nice mathematical properties. Her predecessors had long been hampered by confusion over coordinates. One question that remained unresolved for years, and in fact decades, was whether the gravitational waves that appeared in Einstein's early calculations were genuine physical waves or merely byproducts of the chosen coordinates. That was similar to the debate that ensued over the reality of the singularity that arose in Schwarzschild's solution for a lone spherical mass in a vacuum. Arthur Eddington, a skeptic on both of those possibilities, made the sarcastic comment that gravitational waves propagate "at the speed of thought."[15]

In her 1952 dissertation, Choquet-Bruhat showed that gravitational waves were not mathematical artifacts. In order to make that case, she introduced a new kind of coordinate system—involving so-called harmonic or wave coordinates—in which the differential equations of general relativity assume the desired hyperbolic structure and thus become well-posed and solvable.

One benefit of approaching the problem in this setting is that harmonic coordinates automatically adjust to passing gravitational waves. Instead of thinking about four-dimensional spacetime, which is hard to picture, the physicist Frans Pretorius suggested a simpler analogy. "Imagine dividing a pond into a grid and putting a rubber duck in each section of the grid," he said. "If there's no wave, the ducks just sit there. When a wave moves past, the ducks will bob up and down. Even though the ducks are moving, they maintain a fixed position with respect to the coordinates, and that doesn't change the way you label their positions."[16] For this reason, Choquet-Bruhat's harmonic coordinates offered the ideal framework in which to view, and analyze, gravitational waves.

Choquet-Bruhat made use of another mathematical ploy called 3 + 1 decomposition, which involves taking a four-dimensional spacetime manifold and separating out the time component, so that one instead ends up with a series of separate three-dimensional slices, or surfaces, each representing space at a single moment in time. If you put all these slices together—one after the other, sequentially—you can build up spacetime. That approach might strike some as curious, given that Minkowski went to a lot of trouble to put space and time together. In this case, it became necessary to pry them apart again, by literally pulling time out of the mix, in order to see how space itself is changing from one moment to the next.

Choquet-Bruhat did not invent this 3 + 1 formalism; it was origi- nally developed by the mathematician Georges Darmois in the 1920s, then generalized by André Lichnerowicz (Choquet-Bruhat's PhD advisor) in the 1930s and 1940s, and generalized even further by Choquet-Bruhat herself in a 1948 paper.[17] Nevertheless, although Darmois and Lichnerowicz did recognize that the 3 + 1 strategy could be a useful tool, it was Choquet-Bruhat who first implemented this formalism to prove the existence of a unique solution to the initial value problem in general relativity.

"Her genius was to apply the 3 + 1 split in this case," said the mathematician Martin Lesourd, "and she came at this from a purely mathematical perspective. By doing so, she made general relativity more rigorous and more amenable to mathematical study."[18]

This work alone did not completely prove that the field equations of general relativity were well-posed. Choquet-Bruhat's proof in 1952 is what's called a local existence and uniqueness theorem. *Local* in this context means that if you specify the initial conditions of a sys- tem at a particular moment in time, a single, unique solution will exist for a short while after that, although you don't know if that so- lution will last forever. Even with that caveat, Choquet-Bruhat's work is still considered a huge advance.

In 1969, Choquet-Bruhat teamed up with the mathematical physi- cist Robert Geroch to forge a "global" existence proof, though it was not fully global in that it could not be guaranteed to hold up for all of time. What they did instead was to start with an initial configu- ration, use her local existence and uniqueness proof to evolve a bit forward in time, and keep repeating that process—pushing things ahead, one step at a time, going as far as the equations allowed them to go until a singularity appeared in the solution and forced them to stop.[19]

Which raises the question: How far can mathematicians actually push things? Choquet-Bruhat and Geroch showed that you can carry things into the future for a certain (though not necessarily unlimited) amount of time. But can this evolution continue forever? Is there a setting in which one can assuredly claim to know what will (and won't) happen all the way up to t = infinity?

An important answer came twenty-five years later from the mathematicians Demetrios Christodoulou and Sergiu Klainerman, who evaluated the stability of the simplest possible spacetime—that of flat, empty Minkowski space. Shing-Tung Yau played a role in the formative stages of this project. After Klainerman earned his PhD from New York University in 1978, he began a fellowship at the University of California, Berkeley, and shortly thereafter he visited Yau at Stanford to discuss possible research projects. Klainerman had specialized in nonlinear wave equations—of which the Einstein vacuum equations are a conspicuous example—but he had no interest in general relativity, a sentiment that was prevalent among mathematicians at that time.

Given Klainerman's research background, however, Yau encouraged him to look into the stability of Minkowski space. Yau was personally invested in this question because he and Richard Schoen had recently proved a theorem (to be taken up in Chapter 7), which demonstrated that the mass of an isolated physical system must be positive—or at least nonnegative. He knew from this work that the energy of Minkowski space was zero and could never go below zero, but he did not know if the space itself was dynamically stable. If the space was disturbed in some way—given a "kick," so to speak—it might come out of that exchange with a higher energy. That prospect could not be excluded out of hand. Klainerman eventually joined with Christodoulou, who had also discussed the problem at length with Yau, to prove that would not happen—that the energy of Minkowski space would remain unchanged, locked at zero, even in the face of a perturbation.

The setting of their inquiry was a vacuum, wholly lacking in matter, that might be occasionally disrupted by the passage of weak gravitational waves. An analogous situation would be a lake on a calm, windless day whose otherwise smooth surface is disturbed by someone throwing the occasional stone into it from the shore. When that happens, little waves will form and eventually subside, with the surface becoming smooth again—that is, until the next stone is tossed in. Under such circumstances, the lake is considered stable. The addition of a single stone, or even multiple stones, will not cause things to run amok.

Christodoulou and Klainerman proved something similar with regard to Minkowski spacetime. It too is stable because gravitational waves won't build up and eventually grow into singularities. The solution, however, was far from obvious. Although waves disperse and get weaker in linear theory, they can build up in nonlinear theory. And waves that start out weak don't always remain weak. That was one of the factors that made this problem so challenging.

After spending about seven years working on the problem, Christodoulou and Klainerman constructed a proof in 1993 that was more than 500 pages long.[20] It established an important milestone, according to the mathematician Mihalis Dafermos, "because you can't talk about the stability of black holes if you don't know how to talk about the stability of flat space."[21] And the aforementioned paper on Kerr black hole stability (by Giorgi et al.) did, in fact, follow a general strategy that had been introduced in the Minkowski proof almost thirty years earlier.

The joint work with Klainerman in 1993 represented Christodoulou's first effort to extend geometric analysis—a field pioneered by Yau and various colleagues that combined analysis, a form of calculus, with geometry—taking it from the domain of elliptic equations, in which everything happens in a single instance of time, to the domain of hyperbolic equations, which incorporate the passage of time. Klainerman is also regarded as a leading figure who helped bring hyperbolic equations into the realm of general relativity.

This work by Christodoulou, Klainerman, and others ushered in a "new era in mathematical general relativity based on the marriage of the theory of nonlinear hyperbolic equations and the deep use of global geometry," commented Dafermos,[22] who—as a Harvard undergraduate in the 1990s—was introduced to the subject of geometric analysis by Yau.

In another 1991 paper, Christodoulou delved into the so-called nonlinear gravitational memory effect. The idea is that when a gravitational wave moves through an experimental apparatus, it temporarily displaces test masses that soon come to rest after the waves pass.

Christodoulou showed, however, that the test masses don't return exactly to their original positions; the spacetime geometry retains a memory of the passing gravitational wave, and it is permanently changed by it—an effect big enough that it might be detectable. In this way, he explained, the final flat spacetime remains slightly different from the initial flat spacetime, and that, in turn, is a consequence of the nonlinearity of the Einstein equations.[23] This contrasts with the situation of a lake surface temporarily disturbed by a hurled stone because the lake eventually returns to its original state. Its surface does not retain a lasting memory of the undulations caused by the stone now resting peacefully on the lakebed.

Christodoulou conjectured, though did not prove, that the memory effect—i.e., the permanent displacement of test masses—can be enhanced by other forms of energy, such as electromagnetic radiation, acting in concert with gravitational radiation. The memory effect would be boosted in this way, he speculated, so long as the magnitude of the electromagnetic radiation is not the same in all directions. (This conjecture was proved in 2011 by the mathematicians Lydia Bieri, Po-Ning Chen, and Yau.[24] A 2016 paper in *Physical Review Letters* by the physicist Paul Lasky and colleagues showed how the memory effect might be detected at existing gravitational wave observatories by combining signals from "dozens of nearby events."[25])

In 2007, Christodoulou issued a major, 600-page paper entitled "The Formation of Black Holes in General Relativity." This work builds upon Penrose's contribution in 1965, which proved that the presence of a closed trapped surface implies that a region of spacetime must be closed off from outside observation—in other words, a black hole must be lurking somewhere behind this shrouded veil. The challenge that Christodoulou took up, utilizing equations of the hyperbolic type, was to find out how, in this scenario, trapped surfaces might form in the first place. He showed, following the dictates of general relativity, that trapped surfaces can be created through the "focusing" of gravitational waves—that is, if enough energy in the form of gravitational waves is concentrated into a small enough region. That paper, Dafermos noted, "unleashed a lot of work on similar questions" that is still occupying researchers in mathematical relativity.[26]

It is certainly true that, in the early going, practically all the progress made in general relativity—including that contributing to the understanding of gravitational waves—came through work in mathematics and theoretical physics. But in the 1970s, concurrent advances were being made on the observational front, so that science, once again, began to move forward in the more usual fashion, propelled by a mix of theory and experiment.

A major breakthrough occurred in 1974 when the astronomer Joseph Taylor and his graduate student Russell Hulse discovered the first binary pulsar. A pulsar is a rapidly rotating neutron star that emits a burst of radio waves in periodic pulses. Taylor and Hulse spotted a pair of pulsars in close orbit around each other. Analysis by Taylor and his colleagues over the next four years revealed that this binary system was steadily losing energy, and the amount of energy being lost almost exactly matched (within about one half of a percent) that which general relativity said should be emitted in the form of gravitational waves when two massive bodies circled each other in this fashion.[27] This constituted the first compelling (though indirect) evidence for the existence of gravitational waves, and it also provided added impetus for the development of gravitational wave detectors—work that was already underway.

In 1972, MIT physicist Rainer Weiss invented a gravitational wave detector based on laser interferometry—the premise being that gravitational waves, which simultaneously compress space in one direction and stretch space in a perpendicular direction, could change the way two perpendicular beams of laser light interfere, or combine, with each other at the point where the two beams converge. In 1980, the US National Science Foundation (NSF) funded the construction of a prototype interferometer detector by Weiss and a separate device by a group based at Caltech. A decade later, the agency that oversees the NSF, the National Science Board, approved funding of the Laser Interferometer Gravitational-Wave Observatory (LIGO)—a single observatory consisting of two detectors, one in Louisiana and one in Washington, each spread out over several square miles of land. The detectors were geographically separated from each other, located about 3,000 kilometers (1,865 miles) apart, so that researchers could have greater confidence that a signal they observed was actually due to a gravitational wave rather than to local terrestrial interference of some kind.

Construction at the two sites began in 1994, and initial operations began in 2002. But there was still a big outstanding problem that had to be solved in order to afford LIGO investigators their best chance for success, and it came down to this: Solutions in general relativity for a single mass in spacetime are, as we've seen, hard to come by. Given that, the situation involving two black holes hurtling toward a catastrophic union turns out to be far too complex to solve exactly, or analytically, working with just pen and paper. In fact, the two-body problem—involving any two gravitationally bound masses that may be rather ordinary, having nothing to do with black holes or fiery crashes—cannot be solved exactly at present, and some researchers believe that such a solution is, for theoretical reasons, unattainable.

For problems of this nature, practitioners must instead resort to the methods of numerical relativity, which enlist the power of computers that can make extremely accurate approximations by carrying out literally billions of calculations. A central goal of numerical relativity has been to model a black hole collision and figure out the resulting gravitational waveform, or shape of the waves produced throughout the entire interaction, as well as the amplitude and frequency of those waves. Without this critical input, LIGO investigators would not have known what to look for, nor could they have accurately interpreted what they were seeing.

But the physicists and computer scientists pursuing this objective were held back by a host of obstacles. Some members of this community have since followed a general approach that Pretorius introduced in 2005 in his simulation of a black hole merger, which offered a value for the angular momentum of the combined black hole and suggested that five percent of the system's initial mass would be radiated away as gravitational waves.

A key element of Pretorius's strategy was to adopt a modified version of Choquet-Bruhat's harmonic coordinates. Before then, her work had largely been ignored by the numerical relativity community. Even worse, several conjectures had been circulating that claimed that harmonic coordinates should not be used for modeling gravitational waves—sentiments that had delayed progress in the field. Pretorius, however, wanted to see whether equations that had worked so well in mathematics could be utilized in numerical relativity, too. And he found out they could.

Given that computer codes can't handle singularities, Pretorius also made use of a technique called excision, which allowed him to literally cut a singularity out of a black hole. The strategy is legitimate, he said, because if the singularity is hidden behind an event horizon, "nothing can get outside to pollute the [gravitational wave] signal you are trying to compute."[28]

As has been discussed at length, proving things in general relativity can be very difficult, given the unwieldy nonlinear equations that practitioners must confront. Progress in numerical relativity is therefore welcome and essential, though the results of these efforts do not, and cannot, carry the weight of full mathematical proofs. LIGO researchers are now building up a "library" of solutions to the Einstein equations for collisions of black holes and neutron stars of varying masses. This database, in turn, has helped investigators pick out actual gravitational wave signals and interpret what they see.

And sure enough, something of note was spotted by both LIGO detectors on September 14, 2015. After further analysis and vetting, scientists with the LIGO observatory and its European counterpart, Virgo, announced on February 11, 2016, the first direct observation of gravitational waves—produced, in this case, by the merger of two black holes, an estimated 1.3 billion light-years away, one with the mass of approximately twenty-nine suns and the other with the mass of thirty-six suns. The result was presented at a press conference—which took place in Washington, D.C., that month—almost exactly 100 years after Einstein's prediction that gravitational waves would be unleashed during violent events in the universe.

This discovery, as we've seen, rested on a skillful blending of physics, mathematics, and computer science. Since that time, close to 100 gravitational wave events—involving the mergers of two black holes or two neutron stars or, at least in one case, a black hole and a neutron star—have been detected at LIGO and Virgo.[29] And many more will undoubtedly be seen as more powerful telescopes, based both on the ground and in space, are brought into the search.

This accomplishment, however, goes far beyond the steady accumulation of case studies individually cataloging cosmic collisions of unimaginable fury—important as that may be. The bigger story might be the way in which math and physics, theory and experiment, have synergistically come together here, creating the new field of gravitational wave astronomy and thereby opening a new window onto the universe. Through this recently established portal, scientists are beginning to explore mysteries and phenomena that were previously beyond reach and possibly beyond what our imaginations, unsupported by any experimental input, could have conjured up on their own.

6

An Equation for the Whole Universe

Just as Isaac Newton's interest in gravity started—according to traditional folklore—with an apple falling from a tree, Albert Einstein's interest in that same phenomenon started with his musings about a man falling from a roof. Several years later, Einstein turned his attention to a problem that was bigger in scope: the motions of planets in our solar system, particularly that of Mercury and its idiosyncratic peregrinations around the sun. And in his monumental March 1916 paper, "Foundation of the General Theory of Relativity," he considered how the sun's gravitational field would affect, as well as deflect, light from a distant star.

One year later, Einstein cast his gaze much farther and much wider. He recognized that his theory was not limited to the principles that govern the motion of objects (including beams of light) through the universe, following trajectories dictated by the curvature of spacetime. In a paper entitled "Cosmological Considerations in

the General Theory of Relativity," which he presented at the Prussian
Academy of Sciences in February 1917 (and published in the Academy's proceedings a week later), he explained how the principles of
general relativity could be applied to the universe as a whole, thereby
placing cosmology—a field that had until then relied heavily on
speculation and pontification—onto a much firmer basis. And the
foundations of modern cosmology, which Einstein laid down in that
year, still dominate the field today.

In a letter he wrote in 1953, two years before his death, Einstein
explained the objectives of this endeavor in broad and simple terms:
"We are standing in front of a closed box which we cannot open, and
we try hard to discover what is and is not in it."[1] To get a peek inside
the box representing our universe, scientists in the twentieth century
had a powerful tool at their disposal: mathematics in the form of the
field equations of general relativity. And early in that century, when
observational capabilities were limited at best, mathematics provided
a critical—and sometimes the only—avenue for probing the cosmos.

Describing our vast and possibly infinite universe by a single
equation (or, more accurately, a single *set* of equations) might strike
some as an act of hubris, but Einstein was not one to shy away from
a daunting challenge. Nevertheless, he sensed that he had embarked
on a possibly dubious enterprise, confessing to his friend Paul Ehrenfest, a physicist: "I have perpetuated something again...in gravitation theory, which exposes me to the danger of being committed to
the nuthouse. I hope there are none over there in Leyden so that I can
visit you again safely."[2] He also admitted in a separate letter to the
mathematician and astronomer Willem de Sitter that in this, the first

ever attempt at general relativistic cosmology, he had "erected but a lofty castle in the air.... Whether the model I have formed for myself corresponds to reality is another question."[3]

A fundamental problem facing Einstein's theory, or any theory of gravity for that matter, is the premise that all matter attracts all other matter. And if that were truly the case, what would prevent the universe from collapsing, or even imploding, due to unremitting gravitational attraction? Newton had not provided an answer to that question, but Einstein thought he might have a way to address it. "The conclusion I shall arrive at," he wrote in his 1917 paper, "is that the field equations of gravitation, which I have championed hitherto, still need a slight modification, so that on the basis of the general theory of relativity, those fundamental difficulties may be avoided which have been set forth...as confronting the Newtonian theory."[4]

Einstein sought a model depicting a universe that was static, immobile and unchanging in time, as he—and essentially all of his peers— saw no indication that the universe was expanding or contracting or doing anything other than staying put. To achieve this goal, Einstein made a few assumptions that were, at their roots, all interconnected.

First, in order to treat the universe as a whole, rather than addressing separate parts of it (such as an individual galaxy, star, or black hole), Einstein embraced the principle of homogeneity—the notion that in every direction, and on the largest scales, "the average density of matter...is everywhere the same and different from zero." This supposition that, "in the whole of space," matter and energy are evenly distributed[5] made the problem much more manageable. And it was, in fact, corroborated by future astronomical observations.

He also had to address the problem of calculating the geometry of spacetime in a universe where matter and energy extend to infinity. "I think Einstein showed his greatness in the simple and drastic way in which he disposed of difficulties at infinity," noted Arthur Eddington. "He abolished infinity. He slightly altered his equation so as to make space at great distances bend round until it closed up."[6] Einstein, in other words, fashioned a universe that curled up into a sphere due to the presence of masses, which meant there was no hard edge or boundary that he had to figure into his computations— another simplification that made his task more tractable.

In order to achieve the spherical geometry of a spatially closed universe, he also decided to add a new term, designated by the Greek letter lambda (Λ), to the field equations he had "championed hitherto." This cosmological term or universal constant—now referred to as the cosmological constant—did more than just dispense with the need to determine conditions at an infinitely distant boundary. It also satisfied Einstein's desire to construct through mathematics a static and stationary universe—one in which, as he put it, "the magnitude ('radius') of space is independent of time."[7] A universe, in other words, that conformed to the tranquil picture he and others envisioned.

The Einstein equations in their original 1915 (pre–cosmological constant) form did not describe a static universe but rather one that was constantly in flux—always expanding or contracting, even just a little bit, but never standing completely still. That feature was literally built into, and guaranteed by, the mathematics. But in this instance, Einstein turned away from the math that had carried him so far in his desire to find a solution that fit with the image he held of a placid universe, at rest and in equilibrium. Indeed, the main purpose of the newly added term was to provide a kind of repulsive "antigravity" force, which would counteract gravity's tendency to draw matter together, thereby circumventing the problem of universal collapse that loomed over Newton's theory and threatened his as well. Here's what Einstein's modified field equations looked like with this new addition:

$$R_{ij} - \frac{1}{2} R g_{ij} + \Lambda g_{ij} = T_{ij}$$

or, expressed more simply,

$$G_{ij} + \Lambda g_{ij} = T_{ij}.$$

(This equation is sometimes written with a constant, κ or $-\kappa$—a term that incorporates π, the gravitational constant, G, and the speed of light, c—which is placed directly before the T_{ij} term.)

Given that Einstein inserted lambda on the left-hand side of the equation, it could reasonably be considered a geometrical property of spacetime that's related to curvature. But from a mathematical standpoint, Einstein could just as easily have inserted the term on the right-hand side of the equation (and assigned it an opposite sign), in which case it would have represented a new form of energy permeating the entire fabric of spacetime—the intrinsic energy of empty space, which is how it is normally regarded today.

Einstein never spelled out the precise nature or properties of the cosmological term. He just believed that, whatever it was, it needed to be included in the equations. Yet over the years, Einstein repeatedly wavered over his addition of the cosmological term, just as he'd wavered over his prediction of gravitational waves. Sometimes he affirmed his decision; other times he rued the day he ever introduced it, going so far, reportedly, as to call his amendment to the equations the "greatest blunder" of his career.[8] He elaborated on these sentiments in a 1947 letter, confessing that ever "since I have introduced the term, I always had a bad conscience....I am unable to believe that such an ugly thing should be realized in nature."[9]

But the fact that Einstein expressed such grave doubts about his own creation did not deter others from trying to utilize this same term in their own efforts to advance the field. As the physicist Robbert Dijkgraaf noted, "The great thing of science is that a theory can be smarter than its discoverer and have a life of its own."[10]

With the cosmological term, Einstein had given other researchers an intriguing new variable to play with and manipulate in an attempt to fit their models with the new ideas they were considering and with the observations that were starting to be made. It also opened the door for theorists to experiment with the equations—with or without the newly introduced term—to see what other interpretations might be plausible, other than a universe that neither expands nor contracts.

One could, of course, conjure up all kinds of conceivable universes with wildly different properties. However, among the conceivable universes, only those that satisfied general relativity's field equations could be deemed plausible. And among the possible universes sifted out through mathematics, perhaps one of those might bear a close resemblance to the universe we actually inhabit. Once again,

Einstein had provided the world with a marvelous tool. Ever since, other investigators have had the chance to test it out and see what they could do with it. And it didn't take long to catch a glimpse of other possibilities.

In fact, a solution had already been put forth by Hermann Minkowski in 1908, nearly a decade earlier and about seven years before the promulgation of the original, unadorned equations of general relativity. Those field equations, of course, established an equivalence between the curvature of spacetime and the matter density. Minkowski space is flat, by definition, meaning its curvature is zero, and it has no matter, so its matter density is also zero.

Therefore, Minkowski had constructed a solution—some might say a rather obvious (and almost tautological) one—to the field equations, albeit quite a few years in advance of the latter's formulation.

Minkowski space clearly cannot represent our universe because our universe contains matter, as is abundantly evident. Although this solution is sometimes referred to as trivial, it still represents an important, and in that sense nontrivial, case study in the pantheon of spacetimes.

Nine months after Einstein published the "Cosmological Considerations" paper, his friend de Sitter showed that the modified field equations admitted a different solution—one that pertained to a universe with no matter and a positive cosmological constant. Einstein was not happy, as he didn't believe that his equations should allow for a solution in the absence of matter. He shared his views in a letter to de Sitter: "In my opinion, it would be unsatisfactory if a world without matter were possible." He also told de Sitter that his model contained a spacetime singularity and, therefore, "your solution does not correspond to a physical possibility."[11] He even publicized those criticisms in a 1918 paper, but later conceded—after talking with the mathematician Felix Klein—that his objections were invalid, though he never formally retracted them.[12]

It was initially thought that de Sitter described a static universe, perhaps because in the absence of matter one might suppose that there was nothing left to expand. But in 1923, theoretical work by Hermann Weyl and Arthur Eddington showed that when bits of matter, or test particles, were sprinkled into a de Sitter universe, they would instantly recede from one another.[13]

"Particles initially at rest will scatter," Eddington wrote. "It can easily be verified that there is no such tendency in Einstein's world. A particle placed anywhere will remain at rest.... It is sometimes urged against de Sitter's world that it becomes non-statical as soon as any matter is inserted in it. But this property is perhaps rather in favour of de Sitter's theory than against it."[14]

The de Sitter model, in other words, describes a universe that would undergo expansion everywhere, with the injection of the slightest crumb of matter. And this expansion would be exponential, proceeding at the maximum allowable rate, because in a vacuum there is no gravitationally self-attracting matter around to offset the propulsive (antigravity) push conferred by the cosmological constant.

"Einstein's universe contains matter but no motion, and de Sitter's contains motion but no matter," Eddington explained. "It is clear that the actual universe containing both matter and motion does not correspond exactly to these two models." The good news, Eddington added, is that "we are not now restricted to these two extremes; we have available a whole chain of intermediate solutions between motionless matter and matterless motion from which we can pick out the solution with the right proportion of matter and motion to correspond with what we observe." In other words, de Sitter's solution freed up cosmology, showing its practitioners that they should not be shackled by preconceived notions about a fixed, unchanging background of space. "It was the precursor of the other non-static solutions" to follow, Eddington asserted, and more indeed did come.[15]

First and foremost on this list were the equations advanced in 1922 by the Russian physicist and mathematician Alexander Friedmann, who recognized early on that there was not a unique cosmological solution to the general relativity field equations. Instead, he maintained, there would be a whole family of solutions—portraying universes that were expanding or contracting or alternating between the two—and he was the first person to deliberately, and successfully, set out to find solutions of the non-static, dynamical variety.[16]

"While Einstein and other physicists were looking for ways of avoiding general relativity's prediction of a non-static universe," commented Stephen Hawking, Friedmann was, apparently, the "one man...willing to take general relativity at face value."[17]

Friedmann's equations, which were derived from the general relativity field equations, describe how the universe's size must inevitably change over time, depending on the total matter and energy content and the value of the cosmological term (which could be positive, negative, or zero). Friedmann assumed that the universe had an essentially (though not perfectly) uniform distribution of matter and would look roughly the same in all directions, as well as look the same to all observers viewing from different locales. But unlike Einstein, he did not assume that the universe was unvarying. That conferred a predictive strength to Friedmann's equations that had been lacking before: If you knew the universe's expansion rate and matter content at a given moment, you could try to predict how the universe would evolve over time.

Einstein's initial reaction was to dismiss the results—first alleging that there'd been an error in Friedmann's mathematics and then insisting that the work was physically unrealistic and irrelevant to our universe. But eight months later, Einstein retracted his first claim, acknowledging that he himself had made the calculational error, not Friedmann.[18]

Friedmann's equations, meanwhile, demonstrated that a wide range of cosmological scenarios—a universe that could grow exponentially, shrink, or oscillate—were indeed possible. However, given the dearth of relevant astronomical data, Friedmann's focus was on getting mathematical solutions for expanding and contracting universes under somewhat idealized conditions. He did not consider it worthwhile to try to determine which of these options most closely approximated the universe we reside in. "Our knowledge is totally insufficient for doing numerical calculations and for deciding what kind of world our universe is," he commented in 1922.[19]

That knowledge improved, however, faster than he might have expected. By the late 1920s, important information became available that provided support for a dynamic, expanding universe consistent with Friedmann's models. But Friedmann never saw these results, as he died of typhoid fever in 1925 at the age of thirty-seven. Nevertheless, by rejecting the centuries-old gospel, which held that "the universe was eternal and eternally immutable," Friedmann accomplished "a genuine revolution in science," the mathematicians John Joseph O'Connor and Edmund F. Robertson wrote. "As Copernicus made the Earth go round..., Friedmann made the universe expand."[20]

By 1925, the American astronomer Vesto Slipher had measured radial velocities of forty-one galaxies, which showed how fast they were receding or moving away from us—early clues that the universe might, in fact, be expanding.[21] The astronomer Edwin Hubble, working with Milton Humason, built on Slipher's results, and by 1929 he had established a linear relationship between a galaxy's radial distance and its redshift—or rate of recession: The greater the galaxy's distance, the faster it is moving away from us, which in turn makes its emitted light look redder to observers on Earth. This relation, known as Hubble's law, provided almost incontrovertible proof of an expanding universe, confirming predictions made two years earlier by the Belgian cosmologist and Jesuit priest Georges Lemaître. The case was so compelling that many physicists—including Einstein himself—soon abandoned static models in favor of more dynamic cosmological representations.

Friedmann's exact solution to the gravitational field equations of general relativity was independently derived by Lemaître in the late 1920s and by Howard Robertson and Arthur Walker in the 1930s. The Friedmann–Lemaître–Robertson–Walker (FLRW) model—or the Friedmann–Robertson–Walker (FRW) model, as it's sometimes called—is commonly regarded as the standard model of cosmology, one that describes a universe like our own that is expanding, filled with matter, homogeneous, and isotropic (having the same properties and consistency in every direction).

The original Friedmann equations and subsequent versions—incorporated within the FLRW and FRW models—have proven to be remarkably adaptable. Although they are predicated on a universe in which matter is uniformly spread throughout, they can also explain, for instance, how large-scale celestial structures—such as galaxies, galaxy clusters, and superclusters (which are clusters of galaxy clusters)—can grow from small inhomogeneities in the density of matter. The cosmological term lambda, Λ, included in these equations (and in Einstein's amended 1917 version) has also taken on new meaning in light of evidence that has emerged in recent decades. Specifically, observations of certain kinds of distant stellar explosions (called type 1a supernovae) have indicated that the universe is not only expanding, but that this expansion is proceeding at an accelerated clip. The accelerated expansion, in current parlance, is driven by dark energy—called *dark* because the mechanism behind its inner workings is still inscrutable.

The cosmological constant term in the general relativity equations of Einstein, Friedmann, and others may be dark energy. Or dark energy may turn out to be, as some theorists have it, not a constant at all but rather a quantity—called quintessence—that can vary over time.

Whatever dark energy is, it is already considered to be, by far, the dominant form of energy and mass in the entire universe. Einstein had long wanted to get rid of the cosmological constant, and had often regretted inserting the lambda term into his equations, but now, ironically, it appears to be—in a cosmic sense—the most important thing of all. And it's propelling us, at an ever-increasing pace, to an as-of-yet-uncertain fate.

John Wheeler called the expansion of the universe the "most dramatic prediction that science has ever made."[22] The physicist Abhay Ashtekar added to that statement, claiming that "the most profound predictions" of general relativity owe to the fact that, in this theory, spacetime geometry has become "a dynamical entity, and physics is encoded in its properties.... It is because geometry is unleashed from its fixed, rigid structure that the universe can expand, we can have black holes and ripples of curvature [gravitational waves] can propagate across cosmological distances, carrying away energy and momentum."[23]

The prediction regarding cosmic expansion, as we've seen, was borne out of mathematics and later verified by experiment. But that, of course, was not the end of the story, as it raised some obvious questions: Why was the universe expanding? And what, exactly, was it expanding from?

Lemaître was one of the first scientists to seriously take up this question. If the universe was indeed expanding, he surmised, it must have been smaller in the past. Looking back in time, the entire universe must have been packed into a single particle or "quantum of pure energy"—an entity he called the primeval atom, which was later partitioned into ever smaller pieces that spread out far and wide. This primeval atom, Lemaître wrote in 1931, "was broken into fragments, each fragment into still smaller pieces.... The evolution of the world can be compared to a display of fireworks that has just ended: some few red wisps, ashes, and smoke." From these remnants, he added, "we try to recall the vanishing brilliance of the origin of the worlds."[24] The ideas laid forth by Lemaître provided some of the earliest theoretical underpinnings for what's now known as Big Bang cosmology.

Solid empirical evidence came several decades later when, in 1964, the astronomers Arno Penzias and Robert Wilson discovered a uniform background of radiation left over from the Big Bang—close to what had been predicted in 1948 by the physicists Ralph Alpher and Robert Herman. Subsequent studies of this vestigial light, known as cosmic background radiation, starting in earnest with the Cosmic Background Explorer (COBE) mission in 1989, have led to a much more detailed understanding of the Big Bang and the resultant distribution of matter and growth of structure in the universe.

In 1965, the physicist Robert Dicke and his colleagues confirmed that the mysterious radio signal detected by Penzias and Wilson was, in fact, radiation from the Big Bang that had cooled since the primeval blast to a temperature of about 3 degrees Kelvin. In that same year, Roger Penrose came up with his singularity theorem, and Stephen Hawking began to see a way of tying that theorem to the Big Bang itself, while incorporating some of Friedmann's ideas of an expanding universe. Hawking also placed Lemaître's early notions regarding "the origin of the worlds" onto a much firmer mathematical footing.

"I soon realized that if one reversed the direction of time in Penrose's theorem, so that the collapse became an expansion, the conditions of his theorem would still hold, provided the universe were roughly like a Friedmann model on large scales at the present time,"

Hawking explained. "Penrose's theorem had shown that any collapsing star *must* end in a singularity; the time-reversed argument showed that any Friedmann-like expanding universe *must* have begun with a singularity."[25] This argument was formalized in a joint Hawking–Penrose paper published in 1970, which made the case that the origins of an expanding universe like ours—assuming that general relativity is correct—could inevitably be traced back to an initial singularity.[26]

Hawking later retreated from that categorical statement, however, upon realizing that quantum effects, which assume greater importance on small spatial scales, must also be factored into scenarios like this that involve singularities. He concluded that a yet-to-be-developed theory of quantum gravity would be needed—one that could successfully meld general relativity with quantum theory—in order to fully and accurately describe the earliest moments of our universe and other exotic phenomena.

At present, however, general relativity still stands as the reigning authority on all matters pertaining to gravitation. The source of Hawking's lament—the fact that general relativity and quantum mechanics appear to be incompatible—will be taken up later in this discourse. Before getting there, we will address what may be an even more pressing issue that Einstein's theory has faced in the 100-plus years we've had to contemplate its myriad implications.

= 7 =

The Matter of Mass
(and the Mass of Matter)

Central to Albert Einstein's theory, as we've amply discussed, is the statement that spacetime is curved owing to the presence of mass (and equivalently energy). Nevertheless, it had been an unsettled question for a surprisingly long time as to whether the universe, in a sense, even had the mass needed to induce that curvature. The question, stated more precisely, was whether the total mass in our universe had a positive value. This may seem like a strange problem to worry about, or even consider, but the notion that the universe's total mass is positive was, since the beginning of general relativity, simply assumed to be true. For the theory's equations to be self-consistent, the total mass—as defined through those equations—has to be positive, or at least nonnegative; otherwise, major modifications would be needed. But assuming the mass is positive is not the same as proving it to be so.

Beyond that, an even more fundamental issue had to be addressed: It turns out that even the meaning of mass itself was not fully understood when Einstein's theory was first unveiled, and today that issue has still not been entirely clarified. Geometry, however, has been instrumental in the progress that has been made toward answering this question, as well as addressing the matter of the universe's total mass.

The former question, regarding the positivity of mass, was, until 1979, "one of the major unsolved problems of the theory" of general relativity, in the words of Roger Penrose.[1] Given that physicists had long been unable to solve this problem, Robert Geroch challenged geometers at a 1973 conference at Stanford University to formally prove the conjecture, which stated that in any isolated (and hence closed) physical system, the mass (or energy) must be positive. Since the universe can be thought of as an isolated physical system, this same conjecture could be applied to the universe as a whole. Geroch sought to interest geometers in this problem owing to the fundamental connection between geometry and gravity in general relativity.

A key premise built into this conjecture was the insistence that the local matter density is positive (or at least nonnegative), which is equivalent to a geometric statement—namely that the average curvature at every point in space must also be positive. Maintaining that the matter density is positive on a small scale, in the vicinity of an individual point, is a commonly held assumption in general relativity and an entirely plausible one, given that it is consistent with every credible observation made to date. The question at issue was whether the same positivity condition would apply *globally*, to the *total energy* of an isolated system—including contributions from both matter and gravitation—as viewed from very far away at what's called spatial infinity.

Geroch's lecture made a deep impression on Shing-Tung Yau, who had been invited to the conference to speak on different topics. Yau kept the problem posed by Geroch in mind, but didn't pursue it until several years later. By then, he and his colleague Richard Schoen had completed a series of papers on three-dimensional manifolds of positive scalar curvature. Such a manifold, or spacetime, is called asymptotically flat. Fortuitously, that turned out to be the same setting in which the positive mass conjecture had been framed: a spacetime with some matter in it, and hence some curvature, which becomes flatter and flatter as you move out gradually (and asymptotically) toward infinity. Spacetime never becomes totally flat, like Euclidean space, but you progressively approach flatness as you proceed in the outward direction.

Schoen and Yau, moreover, had just then completed a series of proofs involving minimal surfaces—surfaces like soap bubbles that take up the least area possible while spanning the boundary of a closed loop. They decided to see if minimal surface theory could be successfully brought to bear on the positive mass conjecture—a methodology that had never before been considered by physicists, in part because the mathematical techniques were unfamiliar to them.

The approach did, in fact, work. In their 1979 paper, Schoen and Yau employed a strategy called proof by contradiction: They assumed that the mass is not positive and from there proved the existence of a two-dimensional minimal surface, of infinite extent, lying within a three-dimensional space or manifold. This minimal surface has some unusual properties. But its most important feature, from the point of view of Schoen and Yau, is that this particular surface could not exist in a three-dimensional manifold that has positive scalar curvature—the milieu in which the conjecture was framed. Therein lay the contradiction, implying that their original contention—that mass is not positive—was incorrect. To clarify, Schoen and Yau proved that the mass of such an isolated system is always *nonnegative*: It is positive everywhere except in what's called the ground state—flat Minkowski space—where the mass density is everywhere zero.[2]

Schoen and Yau first proved the conjecture in the *time-symmetric* case: a non-dynamical (so-called spacelike), three-dimensional setting in which time is fixed and nothing changes in time. Of course, our universe is not static, and many physicists—including Stanley Deser, who won the Einstein Medal for his contributions to general relativity—doubted that the proof could be extended into a setting where time flowed freely. Nevertheless, in their second round of attack, Schoen and Yau did prove the more general, *time-varying* case of the conjecture by showing that it could be reduced to the special (static) case they'd already taken care of. A key step in their proof involved solving an equation proposed by Pong Soo Jang, one of Geroch's former students. Although Jang had assumed that his equation was insoluble, Schoen and Yau discovered that—by making one key assumption—a solution could be found.[3]

Schoen used their positive energy theorem to solve all remaining cases of the Yamabe problem—an important problem in mathematics formulated in 1960 by Hidehiko Yamabe, which pertains to the scalar curvature of Riemannian manifolds. The theorem also contributed to the solution of the Riemannian Penrose conjecture—a physics problem advanced by Penrose in 1973 and proven by the mathematicians Gerhard Huisken and Tom Ilmanen about a quarter century later and then proven in a more general setting, shortly thereafter, by Hubert Bray. The conjecture states that the total mass of an asymptotically flat, three-dimensional Riemannian manifold with nonnegative scalar curvature (again the conditions assumed in the positive mass conjecture) must be greater than or equal to the mass contributed by the black holes lying within that space.[4]

A spacetime or universe whose total mass or energy is negative could, under some circumstances, be unstable because there would be no lowest bound on the system's total energy, which could keep dropping indefinitely. That's not a ridiculous concern given that a gravitational field has negative energy, whereas an electric field, for instance, has positive energy (which is how devices like capacitors, commonly used in electric circuits and power supplies, are able to store energy). That difference in sign—the reason that one field has positive energy and the other negative—owes to the fact that opposite electric charges attract each other, whereas two masses, each having the same "sign," also attract. The positive mass theorem alleviates the worry, to some extent, that our universe's energy could be in a state of perpetual decline.

Nevertheless, we cannot conclude that our universe is stable on the basis of this theorem, as it only applies to spacetimes that are asymptotically flat, and we don't know whether our universe meets that standard. Nor do we know what kind of boundary our universe has—or if it has a boundary at all. In addition, there is no commonly accepted definition of the universe's total mass or energy unless we understand the nature of its boundary. Therefore, while the proof of the positive mass conjecture may offer some hope regarding the stability of our universe, that question is far from settled.

In 1981, the physicist Edward Witten proved the positive mass theorem using an approach that was more accessible to physicists because his argument was based on linear equations that were easier to understand than the nonlinear equations employed by Schoen and Yau. Witten's paper also helped make the case for the stability of Minkowski space—"the unique space of lowest energy," he wrote, and as such "there is no state to which it is energetically possible for Minkowski space to decay."[5] In 1990, Witten became the first physicist to win the Fields Medal, arguably the most prestigious prize in mathematics, in part for his work on the positive mass theorem.

Also in 1981, Schoen and Yau built on their prior theorem to prove the positivity of the Bondi mass, which refers to the total mass of an isolated physical system after accounting for losses in mass and energy due to gravitational radiation.[6] In 2017, Schoen and Yau achieved another milestone, extending Witten's work, which did not apply to spacetimes of dimensions greater than four due to a condition he relied on, related to spin, that was central to his proof. Schoen and Yau removed this assumption, completing the proof of the positive mass theorem for a spacetime of dimensions four and above— the only stipulation being that the spacetime or manifold in question has to be asymptotically flat.[7]

Although the positivity of mass was established, mathematically, more than forty years ago, it should be stressed that even today there is not a single, broadly accepted notion of mass itself in general relativity—one that serves as a standard that could be applied in most, if not all, circumstances. Our grasp of the concept, which goes back to the way Einstein initially framed it, is mainly limited to evaluating the mass or energy of an isolated system lying an infinite distance away in an otherwise empty, gravity-free space. One way to think about such a space is to imagine a black hole sitting in the center of an empty sphere. Extend the radius of that sphere toward infinity—where matter is absent and space is almost perfectly flat—and that's where you might want to position a human general relativist, armed with the tools for determining the mass of the compact object inside. Fortunately, one needn't go all the way to infinity to obtain a good estimate of the mass of this hypothetical black hole in the middle of an imaginary sphere. As a practical matter, the radius of this sphere just needs to be sufficiently large compared to the Schwarzschild radius; that would be far enough to shield our dutiful observer from the havoc unleashed locally upon spacetime by the black hole whose mass we seek to ascertain.

The positive mass proofs of Schoen–Yau and Witten made use of the so-called ADM definition of mass, introduced in a series of papers starting in 1959 by the physicists Richard Arnowitt, Stanley Deser, and Charles Misner, who make up the A, D, and M of ADM. They posed the equations of general relativity in terms of an initial value problem, starting with some specified initial conditions and then evolving the equations forward in time. Their definition gained added sharpness because they had adopted the 3 + 1 formalism whose use was pioneered by Yvonne Choquet-Bruhat—pulling the time component out of four-dimensional spacetime in order to obtain a precise determination of the mass at any given moment. Removing time from the mix is entirely permissible in this case, given that mass and energy are conserved quantities, which is another way of saying their values do not fluctuate in time.

While ADM mass is a more precise formulation and, in many ways, an improvement over Einstein's original conception, the approach is again limited to working out the mass of an isolated system as seen from very far away in the space (as opposed to the time) direction—an observational zone called spatial infinity where the geometry of spacetime approaches that of flat Minkowski space.

But what if you want a more detailed view—and a more detailed mass calculation? Suppose the system in question is neither isolated nor infinitely far away. What if, instead, you wanted to figure out the mass confined to a compact region of finite volume? And if, for instance, that region contained not just one but two or more black holes, it would be nice to be able to say something about their individual masses rather than just assessing the mass of the system in aggregate.

Such a long-sought definition is called quasilocal mass—*quasi* rather than entirely local because *local* refers to a single point in spacetime. And to quote Penrose: "There is no *local* definition of mass-energy in general relativity that takes into account" all possible contributions to mass and energy, "including that coming from the gravitational field itself."[8] So, instead, you could try for the next best thing—determining the mass of a patch of space that could be quite small. You might even take a particular region and examine its discrete constituents separately, rather than just seeing the system as a single, distant, indistinct blob and weighing it in whole.

This has been the goal of physicists and mathematicians for many decades. But before trying to define quasilocal mass within a particular system, it was first necessary to establish as a baseline that the total mass is, in fact, positive—a goal, as discussed, that was achieved in 1979. Penrose addressed this very point at a seminar held later that year at the Institute for Advanced Study in Princeton. In view of that recent work, he said, which confirmed the positivity of mass "when measured at spatial infinity, it would seem reasonable to hope that [other] classical problems of general relativity may also be resolved in the not-too-distant future." Such an investment of time could be well spent, Penrose added, given that "general relativity has become, in recent years, an experimentally well-tested theory. So any mathematical results of importance to the classical theory will be assured a permanent place in physics."[9]

The first item on Penrose's wish list was for "some sort of *quasilocal* definition of energy where one does not need to go 'all the way to infinity' in order for the concept to be meaningfully defined."[10] To this end, a number of different formulations have been put forth. Some are more suitable in certain situations, but all the definitions have some drawbacks and none is perfect in every respect.

In 1968, Hawking introduced a comparatively simple definition of quasilocal mass still used by some researchers today. He provided a formula for calculating the mass inside a two-dimensional sphere by determining the extent to which incoming and outgoing light rays—perpendicular to the sphere's surface—would be bent by the matter and energy enclosed within. The virtue of Hawking mass is that it's relatively easy to compute. But there are limitations, as the definition works best in special cases—either in a spacetime that is spherically symmetric, which is an idealization, as nothing in the real world is perfectly round, or in a static spacetime without dynamics.[11]

The Hawking definition suffers from an even more serious drawback: In almost all domains that we know of within Minkowski spacetime, its value turns out to be negative, putting it in direct conflict with the positive (i.e., nonnegative) mass theorem and with our basic understanding of the universe. While Hawking mass has turned out to be a useful concept, in this kind of setting it can't really be interpreted as mass at all, given that the notion of negative mass has no meaning in general relativity.

The Australian mathematician Robert Bartnik offered a new definition of quasilocal mass in 1989, which can be considered a localized version of the positive mass conjecture.[12] Bartnik's idea was to take a region of finite size enclosed by a surface and then, by enveloping it with many layers of surfaces of ever-larger area, extend the finite region to one of infinite size so that its ADM mass can be computed. But the region can be extended in many ways—just as a balloon's surface area can be blown up uniformly or stretched in various directions—each yielding a different ADM mass. The lowest value of ADM mass that can be obtained is, according to Bartnik, the quasilocal mass. "The argument would not have been possible before the positive mass theorem," explained the mathematician Mu-Tao Wang, "because otherwise the mass could have gone to negative infinity," and a minimum mass could never be ascertained.[13]

The paper was popular among some mathematicians because Bartnik's definition was both elegant and concise—described in a paper just three pages long. But there is a practical drawback to this approach. Finding the minimum mass in this way is extremely difficult, according to the mathematician Lan-Hsuan Huang: "It is almost impossible to compute an actual number for the quasilocal mass."[14] The definition of Bartnik mass, Wang added, rests on the validity of conjectures that have not yet been fully established.[15]

In 2003 and 2004, Yau and his math colleague Melissa Liu advanced a notion of quasilocal mass that builds on the 1979 positive mass theorem and on work from the early 1990s by the physicists David Brown and James York. The first step in the Brown and York approach is to wrap the physical system that one wants to measure in a two-dimensional surface. By analyzing the geometry of that surface, and seeing how it is curved in spacetime, one can, in principle, determine the mass enclosed inside.

This method relies on both the *intrinsic* and *extrinsic* geometry of the surface. The intrinsic geometry depends on the distance between two points as measured along curves on that surface, a characteristic that does not change, regardless of how the surface is oriented. If, for instance, you mark two points on a piece of paper, the shortest distance between those points as measured along the surface does not change if the paper is laid out flat or curled up to create a cylinder. An ant crawling between those two points might not be able to tell the difference. But when the surface is viewed from the outside, a flat sheet of paper does look different from that same sheet when rolled up into a tube, and that difference lies strictly in its extrinsic geometry.

Brown and York first measured the geometry of the surface in its natural setting—the spacetime in which the physical system actually resides—and they then measured the same surface in a so-called reference spacetime, for which they selected flat, three-dimensional Euclidean space. With respect to their intrinsic geometry, these two surfaces are indistinguishable. But there can be a difference between the extrinsic geometry in the two settings, Brown and York reasoned, and that difference is due to the gravitational field, which in flat Euclidean space is zero by definition. The gravitational field, in turn, provides a measure of the mass and energy encompassed within the surface—i.e., the quasilocal mass.[16]

The Brown–York definition of quasilocal mass had a few drawbacks, however. It can be positive, even in flat Minkowski spacetime, where it is supposed to be zero. So in that situation, it gives the wrong answer. Moreover, the definition only works in the static, time-symmetric case. And the value of mass that one computes varies depending on the choice of what's called the normal frame. One way to think of the term *normal frame*, suggests Wang, is to imagine an unusual kind of rollercoaster with a thin track that's essentially a wire. At any point along this track, one could draw a tangent line, which is a line that just touches the track at that one point. Now suppose that a rider (facing forward on the rollercoaster) were to stand up at some point on the track and extend their arms straight out. The rider's head would point in a direction perpendicular to the tangent line, as would the rider's arms. That would constitute one choice of a normal frame—a frame in which two directions are selected that are perpendicular to the tangent line and perpendicular to each other as well.

But suppose the rollercoaster rider is attached to the track not by riding in a car but instead via a tube placed between their feet. You might imagine a little Lego man whose feet are glued to a pasta tube strung along a piece of string. Not only can the rider stand straight up, but they can also spin around until they're suspended upside down and then swing back up again to the vertical position. If the rider keeps their arms extended the whole time during that rotation, they will pick out a different normal frame at every angle of their orientation as they loop around from a vertical to an inverted posture and then back to vertical. The way that Brown–York mass works is that, in each of these possible orientations—or in each choice of normal frame—the mass determination could come out differently.[17]

The Liu–Yau definition of mass represented an advance in that it did not suffer from this problem. You could say it was gauge independent or, more specifically, independent of the choice of a normal frame. Moreover, Liu and Yau were the first authors to prove the positivity of quasilocal mass in general. Their work extended the contributions of the mathematicians Yuguang Shi and Luen-Fai Tam who, in a 2002 paper, proved the positivity of Brown–York mass in the time-symmetric case only.[18] Liu–Yau quasilocal mass was not limited in this way, but their definition still suffered from an issue that had also afflicted Brown–York mass: It too remained positive, even in flat Minkowski space (where it should be zero), and that was a problem.

Wang and Yau joined forces in 2008 to address that flaw by making four-dimensional Minkowski spacetime (rather than three-dimensional Euclidean space) their reference spacetime. That approach helps guarantee that quasilocal mass in the Wang–Yau formulation is always positive, except when the background (i.e., physical) spacetime of the surface is also Minkowski spacetime—in which case the mass will be zero, as it should be. This criterion is satisfied because there is no difference in the surface's extrinsic geometry in the physical spacetime and the reference spacetime when both settings are one and the same—Minkowski spacetime. Wang and Yau spelled out other conditions that a definition of quasilocal mass should satisfy and that their definition does meet. If you enclose the physical system in a sphere, its mass should approach the ADM mass as the radius of the sphere goes to infinity. That condition is automatically met by this definition because ADM mass, as discussed earlier, is explicitly defined in the same way.

An additional requirement they insisted on was that "the correct limits need to be obtained when the surface [containing the physical system] converges to a point."[19] This condition is also met by the Wang–Yau definition. The "correct limit" realized at a point—after a procedure called normalization is done to obtain a nonzero limit—would, in fact, be the value of the stress-energy tensor at that point. The stress-energy tensor, the right-hand side of the Einstein equations, describes what happens upon convergence to a point. That's true because the tensor, T_{ij}, specifies the energy—including dark energy and that coming from electromagnetic and other nongravitational fields—as well as momentum and mass at a given point in spacetime. And that is just the kind of information you'd want a definition of quasilocal mass to provide in the vicinity of a point in spacetime.

In their 2008 paper, Wang and Yau expressed their belief that their version "satisfies all the requirements necessary for a valid definition of quasilocal mass, and it is likely to be the unique definition that satisfies all the desired properties."[20] Wang admitted, however, that the approach does have a liability. "Even though our definition is very precise," he said, "it always involves solving several very difficult nonlinear calculations."[21]

In 2015, Wang and Yau used their definition of quasilocal mass to construct, with the help of the mathematician Po-Ning Chen, a definition of quasilocal angular momentum—a challenge that had eluded a solution throughout the first century of general relativity's existence.[22] In fact, when Roger Penrose made a list of major unsolved problems in classical general relativity, it was ranked second. To define quasilocal angular momentum, you first need to be able to define quasilocal mass—which had already been done to their satisfaction—and also define rotation. But in a general four-dimensional space-time, the definition of quasilocal angular momentum depends on the choice of coordinates: With different coordinate choices, one could arrive at different values for the momentum. That was a problem that needed to be addressed.

Chen, Wang, and Yau were able to circumvent this issue by, once again, using Minkowski spacetime as the reference spacetime. Rotation of an object or surface can be readily defined in Minkowski spacetime owing to the presence of rotational symmetries, which in turn is due to its flatness. One way to picture that is to imagine yourself standing in the middle of a crowded city—in New York City's Times Square, for instance. What you see as you turn around a full 360 degrees—in terms of people, vehicles, street signs, buildings, and storefronts—would continually change owing to the lack of rotational symmetry in this particular locale. Alternatively, if you were standing in a perfectly flat, featureless spot in the Great Plains or Mojave Desert, for instance, and swiveled around, every view would look the same due to the presence of rotational symmetry. The same argument regarding symmetry applies to a flat spacetime completely lacking in matter—a fact that Chen, Wang, and Yau exploited to show that, in Minkowski spacetime, the determination of quasilocal angular momentum is not dependent on the choice of coordinates.

They then used a preexisting mathematical theorem to establish a one-to-one correspondence between points on the surface in the natural (physical) spacetime and that same surface when placed in the reference (Minkowski) spacetime. Since you can understand how rotation works in Minkowski spacetime, owing to the symmetries inherent to that milieu, you can use this correspondence to figure out how rotation works on the actual physical surface.

In 2022, Chen, Wang, and Yau, along with the mathematician Ye-Kai Wang, solved another long-standing problem, one dating back to the early 1960s, which relates to the angular momentum carried away by gravitational waves, such as those given off during the merger of two black holes. However, they couldn't just use their prior (2015) definition of quasilocal angular momentum because a precision measurement would be impossible in the immediate vicinity of such an intense event, owing both to the extreme curvature of spacetime there and the tangled pattern of gravitational radiation.[23]

One could make some headway, however, by considering a more familiar situation, which does not involve black holes: that of a large radio broadcast antenna, blasting out radio waves in all directions.[24] If you tried to measure the energy carried by those waves right next to the transmitter, the waves would interfere with each other in complicated ways, making for a difficult measurement. But far from the source, the waves will travel at the speed of light, heading straight or radially outward, and not interacting with each other at all. The intensity—or power transmitted by the waves per unit area—drops off steadily, by a factor of $1/r^2$, with r being the distance from the origin (or, in this case, from the radio antenna). The direction in which light and all forms of electromagnetic radiation propagate is called the null direction, and it sweeps out in a light cone. If you continue to travel along this (null) path as far as you can go, you reach null infinity, a notion Penrose introduced in 1964, which physicists consider the preferred spot from which to observe a black hole.

One could, at least hypothetically, post observers at various spots along the outer edge of this light cone. From there, they could measure the energy of the radio waves and, after pooling their results, determine the total energy transmitted by the antenna. The same approach, in principle, could be used to measure the energy transmitted in the form of gravitational waves, which also travel at the speed of light. The Bondi mass—the mass left behind after gravitational waves carry away energy—is also defined and measured at null infinity. Of course, we can't literally place human observers infinitely far away. Instead, physicists or mathematicians tackling problems of this sort do calculations whose limit lies at null infinity.

That said, there is still a complication that arises when trying to determine the angular momentum carried by gravitational waves. As Abhay Ashtekar noted, "Gravitational waves distort the spacetime in which you do the measurements, and those distortions are not uniform in all directions."[25] This is a consequence of the gravitational wave memory effect: the fact that when gravitational waves travel through spacetime, they leave a permanent imprint.

What that means in the context of our present discussion is that shifts in coordinate systems, which involve moving the origin from one point to another, can lead to different computations of angular momentum. Mass and linear momentum, on the one hand, are not affected by these coordinate shifts because their values are not angle-dependent. Angular momentum, L, on the other hand, is the product (or, actually, the vector product) of the distance from the origin, r, and the linear momentum, p. Owing to the presence of gravitational radiation, as just explained, the value of r is angle-dependent. In other words, lengths or spacetime intervals can be changed by gravitational waves, which is why coordinate ambiguities—or supertranslations, as they're called—may arise.

Because of these ambiguities, Penrose noted in a 1982 paper, "it is hard to see in these circumstances how one can rigorously discuss such questions as the angular momentum carried away by gravitational radiation."[26] (The supertranslation issue also earned a place on his list of the top unsolved problems in general relativity.)

Of course, conserved quantities—like angular momentum—should not vary, or appear to do so, based on how we choose to label things, and that was precisely the situation Chen, Mu-Tao Wang, Ye-Kai Wang, and Yau hoped to rectify. In the aforementioned 2022 paper, they came up with a supertranslation-invariant definition of angular momentum—the first definition ever advanced that was not coordinate-dependent in this way. This definition, in turn, was derived from Chen, Mu-Tao Wang, and Yau's 2015 definition of quasilocal angular momentum. Mu-Tao Wang described the approach as follows: "We first determine quasilocal angular momentum [using the prior definition] at a finite radius and then take the limit as the radius approaches infinity."[27]

The paper, according to the mathematician Demetrios Christodoulou, "essentially settles an important problem in general relativity, the proper definition of angular momentum at future null infinity," sixty years after it was first recognized.[28]

Given that scientists at the LIGO and Virgo observatories have already measured gravitational waves from 100 or so black hole mergers and always seek to extract as much information as possible from these waves, Lydia Bieri said, it was vital "to have unambiguous definitions and a solid mathematical formalism at hand for the very basic concepts involved," which Yau and his collaborators provided.[29]

To elaborate on that point, numerical relativity—upon which LIGO scientists lean heavily—depends on approximations made in order to solve the Einstein equations. And it's important to know, deep down, the actual meaning of the thing or concept that you're trying to approximate. As a practical matter, the observations currently being made in gravitational wave astronomy are not accurate enough for the subtle differences caused by supertranslations to be noticeable, said Vijay Varma, a physicist and member of the LIGO collaboration. "But when the accuracy of our observations gets ten times better, those considerations will become more important." Varma pointed out that improvements of that order are not that far off and should be realized sometime in the next decade.[30]

In the meantime, the average person is not likely to lose sleep worrying about coordinate ambiguities in the definition of angular momentum. Nevertheless, the results obtained by Chen, Mu-Tao Wang, Ye-Kai Wang, and Yau—"the culmination of intricate mathematical investigations over several years," according to Bieri[31]—are not just bits of abstract esoterica. That's because these investigations bear on spectacles, played out on a vast canvas, that we now have the wherewithal to witness, sometimes in full technicolor. Ultimately, there is a benefit to be gained here—that is, if you care about what happens when two intensely gravitating objects (aka black holes) meet in spacetime and energetically conjoin, sending evidence of their fateful union rippling throughout the cosmos from which discerning, and properly equipped, onlookers hope to learn.

8

The Quest for Unification

As we approach the end of this saga, it might be instructive to reflect on where we stand—and on where our explorations of general relativity have taken us so far. On September 14, 2015—almost exactly 100 years after Albert Einstein unveiled the field equations that encapsulate our knowledge of gravitation within a single line—LIGO confirmed the reality of gravitational waves. The 2015 detection also constituted the strongest and most direct piece of evidence obtained until then that black holes, themselves, really exist. Karl Schwarzschild's 1916 solution to the Einstein equations, buttressed by the mathematical theorems of Roy Kerr and Roger Penrose a half century later, has reinforced the notion that the interior of a black hole harbors a singularity: a region in which our conception of spacetime breaks down and where the predictions of Einstein's theory become untrustworthy. In other words, the discovery at LIGO marked an amazing triumph for general relativity, while simultaneously providing almost undeniable indications about the incompleteness—as well as some specific inadequacies—of that same theory.

Physicists contend that a new, broader theory providing a more fundamental description of spacetime is thus needed. This new theory would have to preserve the successes of general relativity (just as general relativity preserved the successes of Newtonian gravity) but also perform well, and reliably, under such extreme circumstances as those found within black hole interiors or, as Stephen Hawking observed, near the Big Bang singularity where general relativity is known to falter. The unified theory that researchers have long sought is often referred to by the term *quantum gravity*, signifying the need to conjoin the laws of quantum mechanics with general relativity—successful theories in their own right that suffer from an unfortunate incompatibility.

Einstein himself took up the search for a more expansive unified theory within just a few years of rolling out his now-famous field equations, although he was motivated to do so for different—though equally compelling—reasons. As we've seen, he was not particularly disturbed by the prospect of objects that contained singularities because he did not believe that such objects could actually exist; he regarded them as mathematical constructs with no basis in physical reality. Yet Einstein was still bothered by the fact that there were—in the early twentieth century—two distinct theories in physics, electromagnetism and general relativity, each of which governed certain aspects of the behavior of objects and particles in our universe, and both of whose reaches extended infinitely far, dropping off according to the same inverse square relation. To him (and, indeed, to many others as well), it would be more natural, as well as more satisfying aesthetically, to have all of this accomplished within one cohesive framework. Moreover, the conservation of charge that holds in electromagnetism had parallels in the conservation of energy and momentum found in general relativity and classical mechanics. Nevertheless, the two theories seemed to operate independently of each other and were even guided by different sets of principles: For instance, where gravity had a geometric interpretation at its root, electromagnetism had not yet been cast in that way.

This dichotomy did not sit well with Einstein, and he vowed to find a single theory that seamlessly stitched together electromagnetism and gravitation, placing them under a common rubric. As he explained in his 1923 Nobel Lecture, "The mind striving after unification of the theory cannot be satisfied that two fields should exist which, by their nature, are quite independent. A mathematically unified field theory is sought in which the gravitational field and the electromagnetic field are interpreted only as different components or manifestations of the same uniform field."[1]

He devoted himself to this endeavor, practically to the exclusion of everything else, for the rest of his life—in the process losing touch with major advancements that were happening around him, particularly within the realm of quantum physics. By most accounts, it was not a triumphant undertaking. "Einstein's quest was primarily a procession of false steps, marked by increasing mathematical complexity, that began with his reacting to the false steps of others," wrote his biographer Walter Isaacson.[2] Einstein, according to the Nobel Prize–winning physicist David Gross, engaged in "a futile search for a unified theory of physics."[3] The American Physical Society summed up Einstein's career trajectory in similar fashion: "After having become famous for several brilliant breakthroughs in physics, including Brownian motion, the photoelectric effect, and the special and general theories of relativity, Albert Einstein spent the last 30 years of his life on a fruitless quest for a way to combine gravity and electromagnetism into a single elegant theory."[4] That was a reasonably accurate rendering, though one could—quite reasonably—take issue with the word *fruitless*.

It is true that Einstein did not succeed in his more than three-decades-long effort. It's also true that a full unification of gravity with the three other fundamental forces now known to exist—electromagnetic, weak, and strong—has not yet been achieved. Nevertheless, a persuasive case can be made that the modern and still-ongoing quest for unification, which Einstein played a key role in instigating, has indeed led to an appreciable harvest of "fruit."

His first paper on a unified field theory was published in 1922, although he had already been thinking about the problem for several years. It would be hard to call that paper a great success. "The time for unification had not yet come," as Abraham Pais later put it.[5] Only two of the four known fundamental forces had been identified up to that point—gravity and electromagnetism. Theories that described the weak and strong nuclear forces were more than a decade off. And there is no foolproof way of unifying that which you do not know and cannot even imagine.

Nevertheless, the goal set by Einstein was definitely worth pursuing, even though his timing was unlucky and, as we now know, a bit premature. "The unification of forces is now widely recognized to be one of the most important tasks in physics," Pais added, "perhaps the most important one."[6] Indeed, as Gross put it, the goal of uniting gravity with the other forces "is the central issue in fundamental physics today." And even if Einstein did not succeed, his influence couldn't have been greater, Gross added. "To all physicists, but especially to those working in speculative areas, Einstein remains an inspiration for his foresight and his unyielding determination and courage."[7]

The challenge Einstein took on has been a major thrust in science—part of a continual drive, dating back to much earlier days, to devise a "system of thought,"[8] as he put it, that could singlehandedly account for a wider range of phenomena. In the 1660s, for example, Isaac Newton began developing a theory of gravity that showed that the terrestrial force, which caused apples to drop to the ground from trees, was the same as the celestial force that keeps the Moon in orbit around Earth and the solar system planets in orbit around the sun. In the late 1700s (as discussed in Chapter 3), Joseph-Louis Lagrange showed that various physical laws, including Newton's laws of motion (from which his theory of gravity was derived), could all spring from a single unifying principle—the action principle. And in the 1860s, James Clerk Maxwell (building on the previous experiments of Michael Faraday) produced a theory of electromagnetism that described the behavior not only of electricity and magnetism but also of light in all its manifestations and frequencies.

Einstein joined in that same tradition, seeking explanations that pertained to an ever more extensive range of phenomena. In fact, he'd already been thinking along these lines in 1917, when he wrote to the mathematician Felix Klein, claiming he had no doubt that the depiction of gravity that emerges from general relativity "will have to yield to another one, for reasons which at present we do not yet surmise. I believe that process of deepening the theory has no limits."[9]

Einstein elaborated on this theme when he spoke, years later, at Columbia University. "We are seeking for the simplest possible system of thought which will bind together the observed facts," he explained. "By the 'simplest' system we do not mean the one which the student will have the least trouble in assimilating, but the one which contains the fewest possible mutually independent postulates or axioms."[10]

In the absence of experimental data to draw upon, Einstein's efforts in this endeavor were instead guided by mathematics—a 180-degree reversal from his earlier point of view when he was fiercely competing against Hilbert to derive the equations of general relativity. He explained his new approach in a lecture given in 1933 at Oxford University: "Our experience up to date justifies us in feeling sure that in Nature is actualized the ideal of mathematical simplicity. It is my conviction that pure mathematical construction enables us to discover the concepts and the laws connecting them which give us the key to the understanding of the phenomena of Nature. Experience can of course guide us..., but the truly creative principle resides in mathematics. In a certain sense, therefore, I hold it is true that pure thought is competent to comprehend the real, as the ancients dreamed."[11]

In a letter to Einstein written four years earlier, Wolfgang Pauli expressed the dismay that he and other physicists shared about Einstein's recently transformed worldview: "All that is left to them now is to congratulate you (or I had better say 'express their condolences'?) on your having gone over to the pure mathematicians."[12]

Another major difference characterized Einstein's work in this area: In contrast to the formulation of general relativity, the major innovations toward establishing a unified field theory were not made primarily by Einstein with help from others, but were rather made by others, with Einstein playing more of a secondary and oftentimes advisory role. The first major strides taken in this direction were made by the mathematicians Hermann Weyl and Theodor Kaluza.

The mathematics of general relativity, as originally formulated, did not provide a ready-made avenue for the geometrization of the electromagnetic force. In a paper that came out in 1918, Weyl showed how Riemannian geometry, within the context of four-dimensional spacetime, might be extended in such a way that it could describe not only gravity but electromagnetism as well. As Weyl saw it, electromagnetism should be regarded as a property of spacetime, which was just how gravitation was seen in general relativity. He attempted to accomplish this melding of forces by incorporating an additional term (called the electromagnetic potential) within the Einstein tensor, G_{ij}, which in ordinary general relativity describes the curvature of spacetime.[13]

Weyl's 1918 paper, according to the physicist Lochlainn O'Raifeartaigh, "showed for the first time how a geometrical significance could be ascribed to the electromagnetic field."[14] Weyl argued that the coordinate invariance of gravitational theory had a counterpart—a scale invariance that was associated with electromagnetism. The notion of scale or gauge invariance, in simple terms, holds that physical laws, and hence physics itself, remain unchanged even when the gauge—the unit of measurement or yardstick—is changed uniformly by a common factor.

The physicist Juan Maldacena offered the following example: Suppose someone wanted to convert US dollars into Argentine pesos at an exchange rate of 3,000 pesos per dollar. Let's further suppose that Argentina then introduces a new denomination of currency (as happened in the late 1980s), the austral, which was worth 1,000 pesos. Someone exchanging a US dollar would get three australs instead of 3,000 pesos. This change in currency is analogous to what physicists call a gauge transformation or gauge symmetry, Maldacena explained, because after this transformation, "nothing changes. Nobody is richer or poorer, and the change offers no new economic opportunities." Maldacena added: "And it does not change anything physical, like the number of bananas you can buy with your salary."[15]

There are, of course, numerous examples in physics where transformations of this sort arise. Currency exchanges are analogous to magnetic potentials that can change from point to point in a magnetic field and are thus related to the variable forces exerted on a charged particle moving within that region. Here's another example: changing the voltage (V) of a system by adding a constant (C) to it has no effect on the electric and magnetic fields. For example, the difference between having an electric potential of 110 volts at one end of a circuit and 100 volts at the other end doesn't change if we add 10 volts to each side. Moreover, if V is a solution to the Maxwell equations of electromagnetism, then $V + C$ is also a solution to those same equations. That's true because V is defined in relation to a reference point, or ground, that is itself arbitrary. Because voltage is not tied to any absolute scale, we classify it as a property that exhibits gauge invariance.

Weyl saw a possible mathematical connection between two different representations of gauge invariance. One was of a geometric nature, which was related to a yardstick or measuring rod whose length could vary from point to point in spacetime, whereas the other gauge was tied to an intrinsic property of the electromagnetic field. This strategy, Weyl surmised, might provide a way to geometrize electromagnetism and, in so doing, link it to the already-geometrized gravity.

But physicists seized upon apparent flaws in this argument. "I just do not believe that the route you have embarked on is the right one, as finely thought-out as it is," Einstein told Weyl in a September 1918 letter, lamenting the fact that "the Lord did not make it easy for us!"[16] Einstein raised a specific objection: The spectrum of electromagnetic radiation emitted by a hydrogen atom would—in Weyl's scenario— depend on the past history of the atom, where *history* in this case refers to the specific path the atom took through spacetime. This was a proposition that experiments carried out on Earth, as well as astronomical observations of distant stars, did not uphold.

Weyl took those comments to heart, telling Einstein a few months later that his criticism "very much disturbs me, of course, since experience has shown that one can rely on your intuition."[17] But rather than giving up, Weyl redoubled his efforts. "It is a tribute to his mathematical insight and self-confidence that he went ahead," claimed the mathematician Michael Atiyah. "The idea was too beautiful to discard."[18]

Weyl, himself, affirmed his conviction that natural laws should express themselves in a mathematically elegant form. "My work always tried to unite the true with the beautiful; but when I had to choose one or the other, I usually chose the beautiful."[19]

By 1929, Weyl had remedied his earlier problem, circumventing Einstein's objection by showing that the movement of, say, a hydrogen atom through spacetime would not affect the spectrum or frequency of the radiation given off but would only affect the *phase* of the emitted electromagnetic waves—the phase having to do with where a particular wave happens to be in its repeating, periodic cycle. That eliminated the previous clash with empirical evidence, enabling Weyl to successfully apply his new gauge theory approach to electromagnetism.[20] The theory he had developed wove together gravitation, electromagnetism, and matter, though its implications were far broader than that: Weyl asserted that gauge theory—or gauge invariance, as he called it—was a general feature of natural laws.[21]

History has subsequently backed up that contention, demonstrating that three of the four known fundamental forces, or interactions, in physics—the electromagnetic, weak, and strong—can be explained through gauge theory. Gravity, however, remains a bit of an outlier in this regard and is best understood, at present, through the equivalence principle.

The discovery of this broader "gauge principle as a fundamental principle of physics was a slow and tortuous process that took more than sixty years," noted O'Raifeartaigh, who broke that process up into three stages: "In the first stage it was shown, mainly by Hermann Weyl, that the traditional gauge invariance of electromagnetism was related to the coordinate invariance of gravitational theory."[22]

"The second stage consisted in generalizing the gauge invariance used in electromagnetism to a form that could be used for the nuclear interactions"—work that began with Weyl's contribution and led to what's now called Yang–Mills gauge theory.[23] That theory, it should be noted, rested largely on the prior work of mathematicians: Weyl, Élie Cartan, Shiing-Shen Chern, André Weil, and others. The physicist Chen Ning Yang (the "Yang" of Yang–Mills who had, in this endeavor, teamed up with the physicist Robert Mills) admitted his unfamiliarity with the theory's mathematical underpinnings when he told Chern that he found it "both thrilling and puzzling [that] you mathematicians dreamed up these concepts out of nowhere." Chern's response was that the development of these concepts in mathematics was not drawn out of thin air; it had, on the contrary, a long and involved history.[24]

In the third stage referred to by O'Raifeartaigh, it was demonstrated that gauge theory could be adapted into a form capable of describing both the weak and strong nuclear interactions.[25] One might also add a fourth stage, as gauge theory has also contributed to more recent advances made toward realizing (though not yet achieving) Einstein's goal of grand unification.

The story did not end there. "Not only is [gauge theory] the framework of modern physics but it is also one of the most novel and exciting areas in modern mathematics," Atiyah maintained. It ties into multiple areas of mathematics, including the geometrical concept of parallel transport and the study of a broad class of geometrical objects called fiber bundles (an important subject in mathematics that we won't have the time or space to delve into here). Atiyah singled out one noteworthy example among many to choose from: "The theory of four-dimensional manifolds due to Simon Donaldson..., which emerged from physics but has turned out to be of profound importance to geometry."[26]

Kaluza, as were those later mathematicians and physicists, was deeply inspired by Weyl's work. However, the work of Kaluza and his followers followed an entirely different track. In a paper he wrote in 1919 that was published two years later, Kaluza stated that although "the residual duality of gravitation and electromagnetism does not diminish [the] beauty of this theory [general relativity], it nevertheless demands its replacement by a totally unified picture." He believed, moreover, that in this paper he was proposing "an even more perfect realization of unification" than that offered within "the profound theory of H. Weyl."[27]

At the heart of Kaluza's thinking were the ten distinct fields or functions required to precisely describe the workings of gravity in four dimensions. To determine the curvature, you need to look at the derivatives (both first and second) of those functions. As we've seen, the force can be represented in the compact mathematical form of a metric tensor, a four-by-four array that has sixteen entries, only ten of which are independent. If you, like Kaluza, wanted to add electromagnetism to the mix, where would you put it? That same four-by-four array cannot accommodate the insertion of electromagnetism because it simply won't fit. Kaluza carved out some additional space by introducing a fifth dimension into this picture, which led him, naturally enough, to a five-by-five array. Gravity's sixteen equations can be nestled inside, leaving room for electromagnetism to be represented by the same, somewhat more spacious, tensor—this one having twenty-five entries, only fifteen of which are independent.

It is perhaps not surprising that the notion of adding an extra dimension, and thereby enlarging the setting in which a unified theory performs its magic, came from a mathematician rather than a physicist. That's because it is common practice now, and was so even a century ago, for mathematicians to think about higher-dimensional and even infinite-dimensional space. Nevertheless, it took a physicist—in this case, Oskar Klein—to fill in some details, both qualitative and quantitative, about the proposed fifth dimension. In 1926, Klein offered an answer to the rather obvious question: If an extra dimension really exists, how come no one has seen it?[28]

Klein suggested that the dimension was extraordinarily compact, curled up into a circle so tiny that it never had been observed. Here's one way to visualize this idea: Imagine a telephone wire, strung tautly and horizontally between two poles. As seen from a distance, it looks like a one-dimensional strand that can only permit movement on a linear path—to the right or the left, but nothing else. But if you were to zoom in and take a close-up view, you would see that the wire's surface is actually a two-dimensional cylinder. A small creature, such as an ant, could move not only along a linear path (from one telephone pole to another) but also in a circular direction, going around the perimeter of the wire, ending up at the same spot it started from.

To describe our spacetime's fifth dimension, Klein invoked a hidden circular direction that, according to his calculations, had to be incredibly tiny: about 10^{-30} centimeters in circumference, putting it close to the so-called Planck length, which—according to current theories of physics—is about as small as you can get. And that's how such a dimension could, in fact, exist without being noticed.

Einstein was intrigued by the possibility of going beyond four dimensions, and—over the years—he personally explored ways of bringing that notion to fruition. "The idea of achieving [unification] with a five-dimensional cylindrical world never occurred to me and...has great appeal for me," he told Kaluza in 1919. "It now all depends on whether your idea will withstand *physical* scrutiny."[29]

But therein lay the rub. Kaluza–Klein theory, as this approach came to be known, ultimately did not bear up under such scrutiny. For one thing, the theory predicted a particle that was never shown to exist. In addition, computations of the ratio of an electron's mass to its charge, based on this theory, were wildly inaccurate.

Nevertheless, the idea was not discarded altogether. On the contrary, it's still quite important, mainly because of the general suggestion made by Kaluza, and elaborated upon by Klein, that some mysteries of our universe may be explained by the presence of dimensions that have so far remained invisible. This is, in fact, a central premise of string theory—a promising but unproven approach to unification—which is based on the notion that spacetime, and hence the universe itself, is a ten- or eleven-dimensional manifold. (There are, by the way, two currently favored versions of string theory—one ten-dimensional and the other, called M theory, which is eleven-dimensional—that physicists believe coexist with each other rather than being competitive. Some string theorists even say that the universe can be both ten-dimensional and eleven-dimensional.) Spacetime, according to this theoretical framework, includes time, the three familiar (and infinitely large) spatial dimensions, plus six or seven miniature spatial dimensions wound up into a tight coil that conceals them from view. "Rather than just postulating the existence of extra dimensions, as had been done by Kaluza, Klein, and their followers, string theory *requires* them," explained the physicist Brian Greene.[30]

The theory—which attempts to combine the two most successful physical theories of the twentieth century, quantum mechanics and general relativity—lies squarely within the realm of quantum gravity. The main innovation here is to replace the point-like objects of particle physics with extended (though still quite miniscule) objects called strings. Forces and particles correspond to different vibrational modes of strings, writhing in higher-dimensional space—a proposition that would never have been taken seriously were it not for the prior efforts of Kaluza and Klein.

String theory requires more than having some extra dimensions for strings to vibrate in. The equations that stem from this theory place serious constraints on the geometrical form these dimensions can take. Their precise size and shape, in other words, have a critical bearing on the kind of universe we live in, determining the physical properties of the particles and forces observed in nature—and even the properties of particles and forces that may exist but have not yet been observed.

In 1984, a group of physicists tried to determine the geometry, or exact shape, of the six hidden dimensions in their attempt to devise a ten-dimensional theory that describes the world we actually inhabit. One of these researchers, Andrew Strominger, contacted Shing-Tung Yau to inquire about the characteristics of spaces that soon came to be known as Calabi–Yau manifolds. That designation stemmed from a conjecture posed in 1954 by the mathematician Eugenio Calabi that Yau proved twenty-three years later. Stated in simple terms, Calabi wanted to know whether certain kinds of manifolds that conform to a general shape or topology could also satisfy very specific, and demanding, geometrical conditions. At the time he presented the conjecture, Calabi assumed that "it had nothing to do with physics. It was strictly geometry."[31]

Yau viewed things differently. Because Calabi's conjecture hinged on Ricci curvature, which can relate to the distribution of matter within a particular space, Yau realized that proving a special case of that conjecture would be equivalent to answering the following question in general relativity: Could there be gravity in a spacetime (or universe) that is totally devoid of matter—a spacetime, in other words, that has zero Ricci curvature? Through his proof, which took many years to complete, Yau ultimately answered that question in the affirmative. In the process, he proved the existence of the multidimensional shapes that were postulated by Calabi and motivated strictly by mathematics.

During his conversation with Strominger, Yau laid out the properties of six-dimensional Calabi–Yau manifolds, and it turns out they had the specific features for which physicists—Philip Candelas, Gary Horowitz, Strominger, and Edward Witten, in particular—were then looking. They needed a way to roll up, or "compactify," the six extra dimensions posited by string theory—thereby making them finite in extent and, in fact, exceedingly small. Calabi–Yau manifolds seemed to be ideally suited for that job.

The Calabi–Yau manifold was embraced by physicists in 1984 and has since became a central component of string theory—about as fundamental to its workings as strings themselves. Just as James Hartle had proclaimed that "gravity is geometry,"[32] one might almost make the broader claim—assuming string theory is correct—that physics is geometry, echoing Plato's much earlier assertion that "God is a geometer."[33] And that's not such an outrageous statement for those who subscribe to the string theory doctrine that—in the ten-dimensional version of the theory—the geometry of the Calabi–Yau manifold dictates the properties of all the particles in nature and all the forces.

But, as mentioned before, string theory has not been substantiated by experiment, and such validation is, by all accounts, going to be very hard to come by. We don't yet know if string theory will turn out to be *the* theory of nature that physicists have long hoped for. Many practitioners regard it, instead, as a step toward the ultimate theory, while still leaving—as the physicist and string theory pioneer Leonard Susskind put it—"a long way to go."[34]

The theory's tentative status, however, should not be taken to mean that it has accomplished nothing so far. String theory has, on the contrary, "taught us many things about how gravity and quantum mechanics fit together," as Susskind said.[35] In 1996, for example, Strominger and his colleague Cumrun Vafa used string theory to provide a detailed picture of a black hole's inner structure. More than two decades earlier, Jacob Bekenstein and Stephen Hawking had demonstrated that a black hole has an unexpectedly and inexplicably high entropy—a term that relates to the number of ways that, on a microscopic level, all the particles and material inside a black hole can possibly be arranged. Strominger and Vafa—bringing the tools of string theory to bear—were able to shed light on that mystery, showing for the first time exactly where that internal complexity came from.[36]

Another prominent example had surfaced six years earlier when Brian Greene (who was then Yau's postdoc) and Ronen Plesser (then Vafa's PhD student) discovered that two distinct Calabi–Yau manifolds, which have different shapes or geometries, nevertheless give rise to the same physics—a phenomenon subsequently dubbed mirror symmetry.[37] This was more than just a weird coincidence. In 1991, a team of four physicists used mirror symmetry to prove a version of a problem first formulated in the late 1800s by the mathematician Hermann Schubert, which roughly translated into calculating the number of spheres that could fit into a six-dimensional Calabi–Yau manifold. The number they arrived at—317,206,375—exactly matched the number derived through more conventional mathematical techniques.[38]

This surprising result gave mathematicians—by taking advantage of the strange correspondence between differently shaped Calabi–Yau manifolds—a new strategy for tackling a range of problems in their field. If it was too difficult to solve a problem by working with one particular Calabi–Yau manifold, they could instead try to approach the problem through its counterpart or mirror pair.

In 1996, Strominger, Yau, and Eric Zaslow provided the first—and perhaps the only—useful explanation of mirror symmetry volunteered so far.[39] According to the so-called SYZ conjecture (named for its three authors), a mirror manifold can be created by taking a six-dimensional Calabi–Yau manifold and breaking it up into two,

three-dimensional submanifolds. These two submanifolds are modified slightly—through some sort of mathematical manipulation—
flipped around, put back together in a different way, and *voila*! A
mirror manifold is born.

The SYZ conjecture has led to a much deeper understanding of
mirror symmetry—a concept that continues to reverberate throughout both the math and physics worlds. This idea helped rejuvenate
the previously dormant field of enumerative geometry, which relates
to counting the number of curves of various types that can fit onto—
or into—different multidimensional surfaces. Mirror symmetry has
also had a sizable impact on algebraic geometry, the study of geometric objects that happen to be solutions to algebraic (and specifically
polynomial) equations: The circle, to pick a simple example, is a solution to equations of the form $x^2 + y^2 = 1$.

Building upon the SYZ conjecture, the mathematicians Mark
Gross and Bernd Siebert developed a fruitful theory of duality in algebraic geometry. It's fair to say that interest in the dualities that can
be found in mathematics and physics—the idea of viewing the same
object or phenomenon through two entirely different frameworks or
lenses—has picked up considerably since the original discovery of
mirror symmetry.

In the process of trying to understand where mirror symmetry
comes from, still a lively area of investigation, mathematicians have
been uncovering new and previously unimagined connections between algebraic geometry and symplectic geometry—the latter being
an approach that, loosely speaking, defines the shape of a space not
as a rigid, unpliable structure but rather as a more flexible entity that

can be characterized by seeing how objects, such as particles or planets, move through it. In situations where this symmetry arises, the same problem could be solved either through algebraic or symplectic geometric methods, whichever approach happens to be easier, thus affording mathematicians a new and potent option to draw upon.

Mathematical spinoffs of string theory continue to unfold. Progress is also being made in physics, although the long-sought unification has yet to be achieved, nor does it appear imminent. But meanwhile, string theory has, for example, provided some of the best descriptions obtained to date of the conditions likely to have prevailed a millionth of a second after the Big Bang when the universe was awash in a hot dense particle soup composed of quarks and gluons. String theory is also being utilized to good effect in condensed matter physics where it has correctly predicted the previously perplexing behavior of electrons in high-temperature superconductors.

It's true that string theory has not fulfilled the lofty forecasts from three decades ago by offering a way to meld general relativity and quantum mechanics into a cohesive package. Nevertheless, there is a silver lining, which might be regarded as the unification of mathematics and physics—or at least a much closer alliance. "Even though string theory has not yet achieved what was initially hoped for," said science historian Peter Galison, "it has opened up new domains of mathematics."[40]

One might perceive a bit of irony attached to that statement: General relativity and the attempt to fuse it with electromagnetism led to the advent of gauge theory and indirectly set the stage for string theory—both of which have fueled widespread and continuing activity in mathematics, even though mathematics and physics are sometimes marked (and marred) by competition. Some mathematicians may regard their work as purer and more rigorous than that of physicists, while physicists may contend that mathematical tinkering is often too abstract, too ethereal, to be of much practical relevance.

Einstein was once a member of the latter camp. Early in his career, he claimed to "not believe in mathematics."[41] He was especially distrustful of the forays (or encroachments) made by mathematicians into his particular areas of research. Einstein initially viewed Minkowski's attempt to geometrize special relativity with suspicion, while comparing Hilbert's mathematics-first, axiomatic approach in formulating a gravitational theory to the labors of a child "unaware of the pitfalls of the real world."[42] Hilbert, of course, had an answer to that, once declaring that "physics is much too hard for physicists."[43]

Einstein, as we've seen, eventually changed his tune. Later in his life, he came to recognize that "a more profound knowledge of the basic principles of physics is tied up with the most intricate mathematical methods"—a revelation that dawned on him "only gradually after years of independent scientific work."[44] That, of course, did not spare him from admonitions from physics colleagues like Pauli who accused him, metaphorically, of shifting his allegiance and crossing over to the dark side.

Still, it would be impossible to argue that new work in physics cannot be inspired by advances in mathematics and vice versa. And the continual exchange and spillover of ideas across disciplinary boundaries is hardly limited to string theory, of course. Just as mirror symmetry and string theory have spurred developments in mathematics, the same is true for general relativity as well. That notion should have been well established by now, in this the book's last chapter. While we may have stressed the contributions of mathematicians toward establishing the general theory of relativity and then unraveling its manifold implications, general relativity has amply returned the favor.

Einstein's encounter, aided by Grossmann, with Riemannian geometry and the tensor calculus of Gregorio Ricci and Tullio Levi-Civita—the mathematical scaffolding upon which he built his theory of general relativity—offers a compelling case in point. General relativity sparked renewed interest in Riemannian geometry, which had been until then "a backwater of mathematics," according to the mathematician Mihalis Dafermos. "The whole reason Riemannian geometry picked up and became an important field in mathematics is undoubtedly because of general relativity."[45]

There's a deeper level to this, too. The mathematician Hung-Hsi Wu observed that when Einstein embraced the curved, higher-dimensional spaces introduced by Bernhard Riemann, he was recognizing something more radical: They "were not just some imaginary things dreamed up by mathematicians," Wu said, "but rather that this was what we needed to understand the universe."[46]

Tensor calculus—first introduced by Ricci as absolute differential calculus and later refined by Levi-Civita—also rose to prominence, in both physics and math, thanks to the gravitational theory of Einstein. "It is not an exaggeration to state that Einstein's general relativity was the killer application for Ricci's [absolute differential calculus]," wrote the science historian Judith Goodstein.[47]

But general relativity did more than just find a purpose for what had been an obscure branch of mathematics. The Ricci tensor, a notion that was conceived about two decades before the birth of general relativity, acquired a new life of its own in mathematics after its appearance in the field equations of 1915. The concept of Ricci flow—a technique pioneered by the mathematician Richard Hamilton that revolves around the Ricci tensor—was taken up and explored by Grigori Perelman. Ricci flow played a central role in Perelman's proof, issued in three installments during 2002 and 2003, of the three-dimensional (and most difficult) case of the century-old Poincaré conjecture—which has provided new insights into what it means to be a sphere in three dimensions.

The spotlight that general relativity focused on four-dimensional spaces eventually led to some major discoveries by geometers and topologists, while also raising some new questions. Interest in this direction was anticipated by the physicist Paul Dirac, who believed there must be something special about four dimensions. In a talk he delivered in 1924 as a Cambridge University graduate student, he laid out his thinking: "The geometrician at present is no more interested in a space of 4 dimensions than space of any other number of dimensions," Dirac said. "There must, however, be some fundamental reason why the actual universe is 4-dimensional, and I feel certain that when the reason is discovered, 4-dimensional space will be of more interest to the geometrician than any other."[48]

Simon Donaldson, who in 1982 began publishing a series of groundbreaking papers on the structure of four-dimensional space, believes that Dirac's statement from the 1920s was "quite prescient.

One of the things that emerges from gauge theories is that four-dimensional spaces have special features," he said. "There are a lot of mathematical things you can do in any dimension. Einstein's equations, for example, work in any dimension, but some things only work in four dimensions."[49] For example, Donaldson noted, in four-dimensional spacetime, "electric fields and magnetic fields look similar, but in other dimensions they are geometrically distinct objects. One is a tensor, and the other is a vector, and you can't really compare them. Four dimensions is a special case in which both are vectors. Symmetries appear there that you don't see in other dimensions."[50]

Donaldson—arguably the world's foremost authority on the subject—cannot yet fully explain, nor can any of his peers, why that is, or why four dimensions are so distinctive. "It's not something we understand in a fundamental way," he added. "It's a mystery to be explored in the future."[51]

Math has been deeply stimulated by general relativity, as well as by attempts to unite general relativity with other areas of physics. There can be—and clearly has been—synergism between the two fields, math and physics, with one feeding off the other. Sometimes this may happen episodically, during bursts of creative interplay, rather than continuously, but it happens all the same.

There have, of course, been prominent scientists on both sides of the "divide" who do not (or did not) put much stock in cooperation between the two fields. The Nobel Prize–winning physicist Richard Feynman, for instance, was not a big fan of such cross-disciplinary ventures. "If all mathematics disappeared today," he said, "physics would be set back exactly one week." To this utterance, Michael Atiyah offered what may have been the perfect rejoinder: "That was the week that God created the world."[52] But as pithy as that remark may be, we (the authors) won't let it stand as the last word here, as we don't want to give the impression that math somehow prevailed in this debate. Instead, we'd suggest that science advances most rapidly, and most assuredly, when we draw on the contributions of people with varied skills and perspectives.

Progress in general relativity—from the very start and, in some sense, even before the field came to an "official" start—has relied on teamwork between mathematicians and physicists, and this is as true today as it was in the early twentieth century. Sometimes this work proceeds separately, with physicists operating in one camp and mathematicians in another. But in those instances when these efforts become intertwined, with one reinforcing the other, we can be confident that our quest to understand the universe—and the enigmatic objects encompassed within—rests on a more secure foundation, even if the foundation itself happens to be the ethereal, four-dimensional amalgam known as spacetime.

Whatever Feynman thought, we need both physics and math to understand the universe, which is surely one of the most noble undertakings of our species. That's a lesson Einstein gradually came around to after some early hesitance on his part regarding the value of mathematics. And we—as in the science community—are still reaping the profits today. Following Einstein's lead—guided by math in one hand and physics in the other—we continue this exalted quest.

Postlude

Wherein the True Mystery Spot Lies

All over Michigan's Upper Peninsula, there are road signs directing visitors toward a tourist attraction: the Mystery Spot. Located about five miles west of the town of St. Ignace (and even closer as the crow flies), the Mystery Spot lies within a 300-foot diameter circle, purportedly discovered by surveyors in the 1950s, where—according to one authoritative source, the *Atlas Obscura*—"gravity does strange things."[1] And other phenomena are routinely witnessed, or said to occur, that seem to defy the laws of physics, nature, and common sense.

But please don't worry that all you've read in the preceding pages has been upended by this 70,000-square-foot patch of northern Michigan and perhaps a few dozen other spots advertising equally weird properties.[2] The facts, documented over the past century, definitively say otherwise. All corroborated observations that have been made throughout the universe by credentialed Earth-based scientists are, in fact, consistent with the theory of general relativity that Albert Einstein set forth in November of 1915.

And in the 100-plus years that have elapsed since Einstein introduced his famous field equations, that theory has been subjected to, and has passed, increasingly stringent tests, including the millions or billions of times each day that we collectively use GPS and our smartphones to navigate the world and communicate with one another.

Within the rarefied halls of academia, active research is still going on. The point isn't just to come up with new tests for the theory. Rather, the field is continually evolving, often bursting ahead in unanticipated ways. It's as if Einstein planted a tree in 1915 that has not stopped growing, steadily becoming more expansive, with its branches reaching out to touch the branches of neighboring trees that were previously unconnected and not known to have any kinship.

A few recent advances on both the experimental and mathematical fronts give a sense—by no means an exhaustive one—of the steady progress that continues to be made. On May 4, 2011, to pick a prominent example, the results of an experiment carried out on a spacecraft called Gravity Probe B—which was launched in 2004 following forty-five years of preparation—were announced at a NASA press conference. After seventeen months of data collection and another five years of data analysis, scientists associated with the project confirmed two key predictions of general relativity. First, they measured the geodetic effect—a minute distortion in spacetime caused by the mass of Earth, which results in the planet's circumference being somewhat smaller than 2π times the radius, thereby exhibiting non-Euclidean behavior—the kind of surprising behavior that Einstein specifically told us not to be surprised by. It was, on the contrary, something to be expected. The second effect that was observed, called frame dragging, shows the extent to which Earth drags spacetime with it as it rotates on its axis. The physicist Clifford Will called the experiment "epic": "One day this will be written up in textbooks as one of the classic experiments in the history of physics."[3]

"Even though it is popular lore [to assume] that Einstein was right," commented Will, "no such book is ever completely closed in science. As we have seen with the 1998 discovery that the universe is accelerating, measuring an effect contrary to established dogma can open the door to a whole new world of understanding, as well as of mystery." While the results of experiments like Gravity Probe B support Einstein's theory, added Will, "this didn't have to be the case. Physicists will never cease testing their basic theories out of curiosity that new physics could exist beyond the 'accepted' picture."[4]

Indeed, the tests have not stopped, but Einstein's theory has withstood every challenge. A 2020 astrophysics paper, authored by an international team of scientists, upheld the equivalence principle to a high degree of precision, long after Einstein had his most exultant realization.[5] Just as Galileo is said to have demonstrated that balls of different weights that are simultaneously dropped from a tower (or released from the top of an inclined plane) will reach the ground at the same time, these researchers demonstrated that two stars of markedly different masses and compositions fall through space under the gravitational influence of a third star with the same acceleration—as seen to an accuracy of two parts per million. "Perhaps more than any previous test, this result indicates that Einstein's most fortunate thought really captures something fundamental about gravity and the inner workings of nature," concluded Paulo Freire of the Max Planck Institute for Radio Astronomy, one of the study's coauthors.[6]

Another major finding, also released in 2020, came after twenty-seven years of observations made at the European Southern Observatory in Chile. Researchers there charted the motion of a star dubbed S2 as it orbited the giant black hole in the center of the Milky Way galaxy. They determined that S2's orbit did indeed precess in close accordance with the predictions of general relativity, just like Mercury's orbit around the sun.[7]

More recently, an international research consortium called NANOGrav, which had spent more than fifteen years monitoring ultrafast-spinning neutron stars (millisecond pulsars) within our galaxy, revealed evidence in June 2023 that our universe may be awash in a background of low-frequency gravitational radiation.[8] The signal that the NANOGrav team members are picking up cannot be traced back to a singular, spectacular collision, they surmise, but instead seems to be emanating from "everything everywhere all at once"—to borrow a phrase from a popular 2022 movie. The collective hum that's been detected, explained Yale astrophysicist Chiara Mingarelli, a member of the collaboration, "could be coming from hundreds of thousands, or possibly a million, overlapping signals from the cosmic merger history of supermassive black hole binaries."[9] Whereas gravitational wave sightings in the past had been isolated phenomena, sometimes separated by weeks or months, they now appear to be a permanent feature of the skies, part of a seemingly continuous and ubiquitous cosmic din.

The experiments keep coming, with no end in sight. Measurements carried out at the CERN Laboratory—as described in a September 28, 2023, paper in *Nature*—showed that hydrogen atoms behave exactly the same in a gravitational field as their antimatter counterparts, antihydrogen atoms, consisting of antiprotons bound to antielectrons (i.e., positrons). This result is consistent with the weak equivalence principle of general relativity, which holds that the motion of bodies under the influence of gravity does not depend on their internal structure. This principle has been validated for matter to extremely high precision, but it had never been directly tested for antimatter before.[10]

At the same time that observations of this sort have been undertaken, and ever more exacting tests administered, advances in mathematical relativity have been unfolding on a parallel track. Progress is still being made, for instance, on the basic definition of mass, particularly quasilocal mass, as well as angular momentum. The issue of stability remains a major topic of research, too. Mathematicians proved in 1986[11] and 1993,[12] respectively, that two vacuum solutions to the Einstein equations—de Sitter spacetime and Minkowski spacetime—were stable, meaning that if these spaces were perturbed in a minor way, they would quickly return to their original state, or close to it. A 2017 paper maintained that a third vacuum solution to those equations, anti–de Sitter spacetime, is unstable.[13] In addition, a 2020 paper (which Shing-Tung Yau coauthored) demonstrated the stability of compactifications of the Calabi–Yau manifolds—a method, somewhat akin to the Kaluza–Klein mechanism, for keeping the six extra dimensions of string theory so small as to be undetectable.[14] This proof came nearly four decades after physicists introduced the concept of compactification as part of what was called the first superstring revolution. There is, however, a major caveat: So far, the stability of the Calabi–Yau compactification as demonstrated in the 2020 paper has only been proven in dimensions greater than twenty-three. Stability has not yet been established in ten dimensions, the actual dimensionality of string theory, so additional work is still called for.

In new 2023 work, a pair of mathematicians completed an existence proof for an infinite family of black holes, configured in a range of exotic shapes, in every dimension beyond four.[15] In that same year, mathematicians also generalized the 1983 Schoen–Yau black hole existence proof to higher dimensions.[16] Mathematical existence, of course, is merely a first step, a prerequisite to actual existence. As to whether any of these fantastical objects can be found in nature is another matter, completely open at the moment.

It is in this realm—the realm of scientific investigation and mathematical rigor, as opposed to the domain of roadside curiosities—that the true mystery spots reside, and it will take some wit to uncover them, as there are no street signs or travel guides to alert us. We have been here before, of course, confronting a world that can often seem baffling and strange. In order to make sense of the cosmos, we'll need to cultivate new tools, tailored for both math and physics, that can help us pry open the "closed box" representing the universe to which Einstein once referred. With these implements in hand, honed through practice, we may have the wherewithal to peer inside that metaphorical box and—to paraphrase the father of relativity again—finally learn what is in it and what is not. And once we've inventoried the box's contents to our satisfaction, we could move on to the even more demanding tasks of figuring out what, if anything, lies outside the box that is supposed to contain all that is knowable, and then determining how that box got there in the first place.

Afterword

Reflections on a Half Century's Work That Was Often Connected, Directly or Indirectly, to General Relativity

Shing-Tung Yau

As I mentioned at the beginning of this rather lengthy disquisition, I was remarkably ignorant of Albert Einstein's gravitational theory when I arrived at the University of California, Berkeley, in the fall of 1969 to begin graduate studies in mathematics. At the time, I was harboring some naïve and wrong-headed notions. I was interested in what I deemed to be "pure mathematics" and nothing else. I believed, moreover, that the purest and deepest things to work on would necessarily be found in the most abstract subject matter possible—the farther removed from the so-called real world, the better. My attitudes quickly changed after my arrival at Berkeley, a sprawling center of intellectual activity where I came into contact with people investigating a broad range of fascinating topics with no rigid demarcations drawn between them. I soaked it all up, auditing as many courses and attending as many lectures as I could fit into a twenty-four-hour day, in addition to the full roster of classes I was officially enrolled in.

Early in 1970, while I was making photocopies in the math office, I had a chance encounter with Arthur Fischer—a lecturer in the department and a newly minted PhD in mathematical physics who had studied under John Wheeler, the renowned general relativist at Princeton. Fischer glanced at the paper I was copying—something I had written during the winter break when the campus was mostly deserted—and he told me that any principle that related the geometry or curvature of an object (such as a surface or manifold) to its general shape or topology could be of importance to physics.

I admit I was intrigued by Fischer's unsolicited remarks, but I was also a bit wary. He looked to me like kind of a hippie—part of a subculture I had essentially no contact with in my decidedly unhip Hong Kong days—so I wasn't sure how much stock to put in his words. Nevertheless, Fischer was teaching a course on general relativity in the spring semester, and I attended some of his lectures, hoping to finally learn something about this topic. It was there, while sitting in on these sessions, that I grasped the fact that curvature is central to the workings of gravity as well as to the geometry I was already studying. I also learned, to my surprise, that geometry plays a vital role in physics that extends far beyond its importance to general relativity alone.

During one of Fischer's lectures, my mind wandered, and I began thinking about gravity in spacetime regions devoid of matter. That led me, in turn, to the Calabi conjecture, setting me on a path that eventually resulted in my rendezvous with string theorists and the attempt—at least in the eyes of some enthusiasts in those heady days—to construct a "theory of everything." That theory has since achieved many important successes, in both physics and math, but it's still a long way from reaching "everything."

A major geometry conference held at Stanford in 1973—where the physicist Robert Geroch tried to lure mathematicians into taking on the long-unsettled positive mass conjecture—drew me further into the orbit of general relativity from which I've never escaped, nor did I ever have the desire to. That's when it finally dawned on me that mathematicians, and perhaps even myself, might be in a position to contribute in a material way to key questions in physics. The problem raised by Geroch stayed with me for a long time, and later in the decade—after having amassed the necessary mathematical skills—I took on that challenge with my friend and colleague Richard Schoen.

Since then, I have often gravitated toward problems that lie at the interface between math and physics—a setting that can be quite stimulating. Physicists can supply mathematicians with ideas we've never thought of before. We can then reframe those ideas in exacting mathematical terms and, hopefully, prove things that will hold for all of eternity—tasks that physicists are often uninterested in, or incapable of, carrying out.

When it comes to general relativity, I, of course, am just one cog in the vast machine of researchers—physicists, astronomers, cosmologists, mathematicians, computer scientists, space engineers, and so forth—who have contributed to a growing body of knowledge over the past century and more. The field has advanced dramatically, just over the course of my career alone. When I started to think about general relativity fifty-some years ago, almost no one believed in black holes. And if you took black holes seriously—say by trying to prove their existence mathematically (which is something that Schoen and I undertook)—many would have dismissed you as crazy.

By now, the empirical evidence for black holes is all but incontrovertible. And for theoretical physicists and mathematicians alike, these objects—once confined to the realm of science fiction—now offer some of the main testing grounds for probing the limits of general relativity and for assessing the viability of various approaches to quantum gravity. In addition, there are many critical issues related to black holes that mathematicians can and should weigh in on: The no-hair theorem, cosmic censorship, and a full resolution of Kerr black hole stability (not only in cases of low angular momentum) are just a few noteworthy examples.

There's still plenty to keep mathematicians busy, and I continue to work on problems of a mathematical bent that bear on general relativity. I hope to keep on doing so as long as I can be productive, though, at some point, my part in these proceedings may be relegated to that of an occasional advisor and spectator. But what a spectacle to behold! And it is amazing for me to consider that problems tied to general relativity have captivated me for half a century and entranced others, collectively, for well over a century.

Enthusiasm for this enterprise, remarkably, has not diminished in a discernible way and, if anything, seems to be increasing. Even if one of the revered scholars in the field once called Einstein a "lazy dog," that particular lazy dog started something big. And the exciting thing about that for me—and I imagine for many other participants and observers—is that we have no idea where this ferment of sustained activity will ultimately take us. Fasten your seatbelts, passengers of

Spaceship Earth. The journey through spacetime should prove to be an interesting, unpredictable, and inescapably bumpy ride.

Ode to Geometry

The bounty of heaven, so vast and beautiful.
Who could not but marvel at its miraculous display?

Theories conceived by thinkers of the past still abound.
While these sages may be gone, their methods remain sound.

Form and beauty meet so agreeably, converging in a perfect way.
Just as mind and substance meld together, as oft they may.

A new century has dawned, bringing fresh hopes and dreams.
Summoning our collective strength, we seek truth by all means.

With telescopes mounted on hilltops or strapped to orbital craft,
the attempt to fathom the Big Bang is no longer considered daft.

The inquiry can be seen as nothing more, and nothing less,
than a probe into the origins of everything and what has come from that.

Apples falling to the ground and planets tracing ellipses around
*   the sun—*
all reduce to the union of space and time and the myriad ways
*   it can bend.*

Serenity lies at a distance, asymptotically, where all is flat and calm.
At the other extreme lies the violent, infinite warping of ravenous
   black holes.

These seemingly inscrutable objects, enigmatic, shrouded in darkness,
reveal their secrets, in time, through the unyielding power of
   geometry.

Honed over the centuries and enduring for millennia,
these tools, and their associated theorems, have never let us down.

Truth is elusive, defying the best minds that history has produced.
Yet a simple mathematical proof can guide us, inexorably, to the
   eternal.

Notes

Prelude

1. G. B. M., "Apollonius of Perga," *Nature* 54 (August 6, 1896), 314–315.
2. Robbert Dijkgraaf, "The Two Forms of Mathematical Beauty," *Quanta*, June 16, 2020, https://www.quantamagazine.org/how-is-math-beautiful-20200616/.
3. Chen Ning Yang, "Albert Einstein: Opportunity and Perception," *International Journal of Modern Physics A* 21:15 (2006), 3031–3038.
4. Robbert Dijkgraaf, "Without Albert Einstein, We'd All Be Lost," *Wall Street Journal*, November 5, 2015.
5. Albert Einstein, "The Mechanics of Newton and Their Influence on the Development of Theoretical Physics," in *Ideas and Opinions* (New York: Wing Books, 1954), 253.
6. Dan Falk, "A Debate over the Physics of Time," *Quanta*, July 19, 2016, https://www.quantamagazine.org/a-debate-over-the-physics-of-time-20160719/.

Chapter 1: Falling Objects, Shifting Paradigms

1. Charles W. Misner, Kip S. Thorne, and John Archibald Wheeler, *Gravitation* (Princeton, NJ: Princeton University Press, 2017), 3.
2. R. G. Keesing, "The History of Newton's Apple Tree," *Contemporary Physics* 39:5 (1998), 377–395.
3. Arthur Rosenthal, "The History of Calculus," *American Mathematical Monthly* 58:2 (February 1951), 75–86.
4. Stephen Hawking, *A Brief History of Time* (New York: Bantam Books, 1988), 181.
5. Ofer Gal and Raz Chen-Morris, "The Archaeology of the Inverse Square Law (1)," *History of Science* 43:4 (2005), 391–414.
6. D. T. Whiteside, "Newton's Marvellous Year: 1666 and All That," *Notes and Records of the Royal Society of London* 21:1 (June 1966), 32–41.

7. Stephen Hawking, "Newton's *Principia*," in Stephen Hawking and Werner Israel, eds., *Three Hundred Years of Gravitation* (Cambridge, UK: Cambridge University Press, 1987), 1.

8. Steven Weinberg, "Newtonianism and Today's Physics," in Stephen Hawking and Werner Israel, eds., *Three Hundred Years of Gravitation* (Cambridge, UK: Cambridge University Press, 1987), 7.

9. W. David Woods and Frank O'Brien, "Apollo 8: Day 5: The Green Team," *Apollo Flight Journal*, updated February 27, 2021, https://history.nasa.gov/afj /ap08fj/24day5_green.html.

10. "Original Letter from Isaac Newton to Richard Bentley, Dated 17 January 1692/3," *The Newton Project*, October 2007, http://www.newtonproject.ox.ac .uk/view/texts/normalized/THEM00255.

11. George Smith, "Newton's Philosophiae Naturalis Principia Mathematica," *Stanford Encyclopedia of Philosophy*, December 20, 2007, https://plato.stanford .edu/entries/newton-principia/#OveImpWor.

12. Michael Seeds, *The Solar System*, Sixth Edition (Belmont, CA: Thomson/ Brooks Cole, 2008), 94.

13. Steven Weinberg, *Gravitation and Cosmology* (New York: John Wiley & Sons, 1972), 14.

14. David Bodanis, *E=mc²: A Biography of the World's Most Famous Equation* (New York: Berkley Publishing Group, 2005), 5.

15. Albert Einstein, "*Über einen die Erzeugung und Verwandlung des Lichtes betreffenden heuristischen Gesichtspunkt*" (On a Heuristic Point of View About the Creation and Conversion of Light), *Annalen der Physik* 322:6 (1905), 132–148.

16. Albert Einstein, "*Über die von der molekularkinetischen Theorie der Wärme geforderte Bewegung von in ruhenden Flüssigkeiten suspendierten Teilchen*" (Investigations on the Theory of Brownian Motion), *Annalen der Physik* 322:8 (1905), 549–560.

17. Albert Einstein, "*Zur Elektrodynamik bewegter Körper*" (On the Electrodynamics of Moving Bodies), *Annalen der Physik* 322:10 (1905), 891–921.

18. Albert Einstein, "*Ist die Trägheit eines Körpers von seinem Energieinhalt abhängig?*" (Does the Inertia of a Body Depend Upon Its Energy Content?), *Annalen der Physik* 323:13 (1905), 639–641.

19. Albert Einstein, *Autobiographical Notes*, ed. Paul Arthur Schilpp (Peru, IL: Open Court Publishing Company, 1999), 49–51.

20. Albert Einstein, "What Is the Theory of Relativity?," in *Ideas and Opinions* (New York: Wing Books, 1954), 229–230.

21. Ibid.

22. Ibid.

23. Galileo Gallilei, *Dialogue Concerning the Two Chief World Systems*, trans. Stillman Drake (New York: Modern Library, 2001), 216–217.

24. Albert Einstein, "How I Created the Theory of Relativity," trans. Yoshimasa A. Ono, *Physics Today* 35·8 (August 1982), 47.

25. Ibid.

26. Anna M. Nobili, "Testing the Weak Equivalence Principle with Macroscopic Proof Masses on Ground and in Space: A Brief Review," *International Journal of Modern Physics: Conference Series* 30 (May 2014), 1460254.

27. Ivan T. Todorov, "Einstein and Hilbert: The Creation of General Relativity," arXiv:physics/0504179v1, April 25, 2005.

28. John Gribbin, *Einstein's Masterwork: 1915 and the General Theory of Relativity* (New York: Pegasus Books, 2016), 16.

29. Vesselin Petkov, ed., *Space and Time: Minkowski's Papers on Relativity* (Montreal: Minkowski Institute Press, 2012), 55 and 111.

30. Anthony Zee, *Einstein Gravity in a Nutshell* (Princeton, NJ: Princeton University Press, 2013), 175.

31. Richard Garfinkle and David Garfinkle, *X Marks the Spot* (Boca Raton, FL: CRC Press, 2021).

32. Mu-Tao Wang (Columbia University), interview with the author, May 5, 2019.

33. Peter Galison, "Minkowski's Space-Time: From Visual Thinking to the Absolute World," *Historical Studies in the Physical Sciences* 10 (1979), 95.

34. Abraham Pais, *Subtle Is the Lord: The Science and the Life of Albert Einstein* (New York: Oxford University Press, 2008), 152.

35. Petkov, *Space and Time*, 2.

36. Matsatsugu Sei Suzuki, "Minkowski Space-Time Diagram in the Special Relativity," Lecture Notes on Modern Physics, Department of Physics, SUNY at Binghamton, January 13, 2012.

37. C. Lanczos, "Einstein's Path from Special to General Relativity," in L. O'Raifeartaigh, ed., *General Relativity: Papers in Honor of J. L. Synge* (New York: Oxford University Press, 1972), 5–19.

38. Jürgen Renn and Hanoch Gutfreund, *Einstein on Einstein* (Princeton, NJ: Princeton University Press, 2020), 84.

39. Albert Einstein, "Minkowski's Four-Dimensional Space," trans. Robert W. Lawson, in *Relativity: The Special and the General Theory* (New York: Crown, 1961); reprinted in Ann M. Hentschel, trans., *The Collected Papers of Albert*

Einstein, vol. 6, *The Berlin Years: Writings, 1914–1917*, English translation supplement (Princeton, NJ: Princeton University Press, 1997), 306–308.

40. Leo Corry, "Einstein Meets Hilbert on the Way to General Relativity," presented at Harvard Black Hole Initiative, October 12, 2020.

41. Leo Corry (Tel Aviv University), email to the author, May 21, 2021.

42. Anthony Zee, *On Gravity: A Brief Tour of a Weighty Subject* (Princeton, NJ: Princeton University Press, 2018), 62.

43. Einstein, "What Is the Theory of Relativity?," 231.

44. Judith R. Goodstein, *Einstein's Italian Mathematicians: Ricci, Levi-Civita, and the Birth of General Relativity* (Providence, RI: American Mathematical Society, 2018), 90.

45. Michel Janssen and Jürgen Renn, "Einstein Was No Lone Genius," *Nature* 527 (November 19, 2015), 298.

Chapter 2: Finding a General Path Forward

1. Bernhard Riemann, *On the Hypotheses Which Lie at the Bases of Geometry*, ed. Jürgen Jost (Switzerland: Birkhauser, 2016), v.

2. Gerrit van Dijk and Masato Wakayama, eds., *Casimir Force, Casimir Operators and the Riemann Hypothesis: Mathematics for Innovation in Industry and Science* (Berlin: De Gruyter, 2010), v.

3. Riemann, *On the Hypotheses Which Lie*, v.

4. Steven Weinberg, *Gravitation and Cosmology* (New York: John Wiley & Sons, 1972), 5.

5. Ruth Farwell and Christopher Knee, "The Missing Link: Riemann's 'Commentatio,' Differential Geometry and Tensor Analysis," *Historia Mathematica* 17 (1990), 224.

6. Bernhard Riemann, *Bernhard Riemann, Collected Papers*, trans. R. Baker, C. Cristenson, and H. Order (Heber City, UT: Kendrick Press, 2004), 257–270.

7. Marcia Bartusiak, *Einstein's Unfinished Symphony: Listening to the Sounds of Space-Time* (New Haven, CT: Yale University Press, 2017), 24–25.

8. Albert Einstein, "How I Created the Theory of Relativity," trans. Yoshimasa A. Ono, *Physics Today* 35:8 (August 1982), 47.

9. James Overduin, "The Experimental Verdict on Spacetime from Gravity Probe B," in Vesselin Petkov, ed., *Space, Time, and Spacetime: Physical and Philosophical Implications of Minkowski's Unification of Space and Time* (Berlin: Springer, 2010), 31.

10. Abraham Pais, *Subtle Is the Lord: The Science and the Life of Albert Einstein* (New York: Oxford University Press, 2008), 213.

11. Ibid., 210.

12. David E. Rowe, "Book Review: Einstein's Italian Mathematicians: Ricci, Levi-Civita, and the Birth of General Relativity," *Notices of the American Mathematical Society* 166 (October 2019), 1478.

13. E. B. Christoffel, "*Ueber die Transformation der homogenen Differentialausdrücke zweiten Grades,*" *Journal für die Reine und Angewandte Mathematik* 70 (1869), 46–70.

14. Galina Weinstein, "Genesis of General Relativity," arXiv:1204.3386, April 16, 2012.

15. Lewis Pyenson, "Einstein's Education: Mathematics and the Laws of Nature," *Isis* 71:3 (September 1980), 419.

16. Rowe, "Einstein's Italian Mathematicians," 1481.

17. Albert Einstein, "Outline of a Generalized Theory of Relativity and of a Theory of Gravitation (I. Physical Part)," *Zeitschrift für Mathematik und Physik* 62 (1914), 225–261.

18. Ibid.

19. Michel Janssen and Jürgen Renn, "Arch and Scaffold: How Einstein Found His Field Equations," *Physics Today* 68:11 (November 2015), 34.

20. Albert Einstein, "Notes on the Origin of the General Theory of Relativity," in *Ideas and Opinions* (New York: Wing Books, 1954), 289.

21. John Norton, "How Einstein Found His Field Equations: 1912–1915," *Historical Studies in the Physical Sciences* 14:2 (1984), 253.

22. John Earman and Clark Glymour, "Lost in the Tensors: Einstein's Struggles with Covariance Principles 1912–1916," *Studies in History and Philosophy of Science* 9:4 (1978), 260.

23. Albert Einstein and Marcel Grossmann, "Covariance Properties of the Field Equations of the Theory of Gravitation Based on the General Theory of Relativity," *Zeitschrift für Mathematik und Physik* 63 (1914), 215–225.

Chapter 3: The Magnus Opus

1. Ann M. Hentschel, trans., *The Collected Papers of Albert Einstein*, vol. 8, *The Berlin Years: Correspondence, 1914–1918*, English translation supplement (Princeton, NJ: Princeton University Press, 1997), Document 60, 71.

2. Galina Weinstein, "Einstein the Stubborn: Correspondence Between Einstein and Levi-Civita," arXiv:1202:4305, January 31, 2012.

3. David E. Rowe, "Book Review: Einstein's Italian Mathematicians: Ricci, Levi-Civita, and the Birth of General Relativity," *Notices of the American Mathematical Society* 166 (October 2019), 1481.

4. Francesco dell'Isola, Emilio Barchiesi, and Luca Placidi, "Levi-Civita, Tullio," in H. Altenbach and A. Öchsner, eds., *Encyclopedia of Continuum Mechanics* (Berlin: Springer, 2019), 1–11.

5. Abraham Pais, *Subtle Is the Lord: The Science and the Life of Albert Einstein* (New York: Oxford University Press, 2008), 259.

6. Ivan T. Todorov, "Einstein and Hilbert: The Creation of General Relativity," arXiv:physics/0504179, April 25, 2005.

7. Jürgen Renn and Matthias Schemmel, eds., *The Genesis of General Relativity*, vol. 4, *Gravitation in the Twilight of Classical Physics: The Promise of Mathematics* (Dordrecht: Springer, 2007), 1003.

8. Constance Reid, *Hilbert-Courant* (New York: Springer-Verlag, 1986), 127.

9. Leo Corry, "The Influence of David Hilbert and Hermann Minkowski on Einstein's Views over the Interrelation Between Physics and Mathematics," *Endeavor* 22:3 (1998), 95–97.

10. Pais, *Subtle Is the Lord*, 259.

11. Todorov, "Einstein and Hilbert."

12. Albert Einstein, "Explanation of the Perihelion Motion of Mercury from the General Theory of Relativity," *Sitzungsberichte der Königlich Preußischen Akademie der Wissenschaften zu Berlin* (submitted November 18, 1915), 831–839.

13. Derek Raine, "Review: Mercury's Perihelion from Le Verrier to Einstein," *British Journal for the Philosophy of Science*, 35:2 (June 1984), 188.

14. Jürgen Renn and John Stachel, "Hilbert's Foundation of Physics: From a Theory of Everything to a Constituent of General Relativity," Max Planck Institute for the History of Science Preprint 118, 1999.

15. Renn and Schemmel, *The Genesis of General Relativity*, vol. 4, 1015.

16. Leo Corry, Jürgen Renn, and John Stachel, "Belated Decision in the Hilbert-Einstein Priority Dispute," *Science* 278 (November 14, 1997), 1270–1273.

17. Martin Harwit, *In Search of the True Universe: The Tools, Shaping, and Cost of Cosmological Thought* (New York: Cambridge University Press, 2013), 35.

18. Kip Thorne, *Black Holes and Time Warps: Einstein's Outrageous Legacy* (New York: W. W. Norton, 1994), 117–119.

19. John Earman and Clark Glymour, "Einstein and Hilbert: Two Months in the History of General Relativity," *Archive for History of Exact Sciences* 19:3 (1978), 307.

20. John Norton, "How Einstein Found His Field Equations," *Historical Studies in the Physical Sciences* 14:2 (1984), 263.

21. Dieter Ebner, "How Hilbert Has Found the Einstein Equations Before Einstein and Forgeries of Hilbert's Page Proofs," arXiv:physics/0610154, October 19, 2006.

22. Pais, *Subtle Is the Lord*, 275–276,

23. Ibid.

24. Fabio Toscano, "Luigi Bianchi, Gregorio Ricci Curbastro e la scoperta delle identita di Bianchi," in *Atti Del XX Congresso Nazionale Di Storia Della Fisica E Dell'astronomia* (*Proceedings of the XX National Congress of the History of Physics and Astronomy*) (Naples: CUEN, 2001), 353–370.

25. Jürgen Neffe, *Einstein: A Biography* (New York: Farrar, Straus and Giroux, 2007), 206.

26. Albert Einstein, "Notes on the Origin of the General Theory of Relativity," in *Ideas and Opinions* (New York: Wing Books, 1954), 289.

27. Tilman Sauer, "Marcel Grossmann and His Contribution to the General Theory of Relativity," in Robert T. Jantzen and Kjell Rosquist, eds., *Proceedings of the Thirteenth Marcel Grossmann Meeting on General Relativity* (Singapore: World Scientific, 2015), 487.

28. Tilman Sauer, "Marcel Grossmann and His Contributions to the General Theory of Relativity," arXiv:1312.4068, April 22, 2014, 32, 35–36.

29. Alberto Rojo and Anthony Block, *The Principle of Least Action: History and Physics* (Cambridge, UK: Cambridge University Press, 2018), 7.

30. Cumrun Vafa, *Puzzles to Unravel the Universe* (Middleton, DE: self-published, 2020), 23–24.

31. David Garfinkle (Oakland University), interview with the author, June 8, 2021.

32. Katherine Brading, "How It All Began: The Puzzle That Led to Noether's Theorems," presented at Boston University, October 19, 2018.

33. Yvette Kosmann-Schwarzbach, *The Noether Theorems* (New York: Springer-Verlag, 2011), 45–46.

34. Emmy Noether, "Invariant Variational Problems," *Nachrichten der Königlichen Gesellschaft der Wissenschaften zu Göttingen, Mathematisch-Physikalische Klasse* (1918), 235–257.

35. This analogy was suggested by the physicist Burkhard Schwab in a June 15, 2021, email to the author.

36. Chris Quigg, "Colloquium: A Century of Noether's Theorem," technical report FERMILAB-PUB–19-059-T, arXiv:1902.01989, July 9, 2019.

37. Ruth Gregory, "Celebrating the Life and Legacy of Emmy Noether," presented at the Perimeter Institute for Theoretical Physics, June 22, 2015.

38. David E. Rowe, "Emmy Noether on Energy Conservation in General Relativity," arXiv:1912.03269, December 4, 2019.

39. Albert Einstein, "Hamilton's Principle and the General Theory of Relativity," in Ann M. Hentschel, trans., *The Collected Papers of Albert Einstein*, vol. 6, *The Berlin Years: Writings, 1914–1917*, English translation supplement (Princeton, NJ: Princeton University Press, 1997), 240.

40. Hanoch Gutfreund, "Relatively Speaking—Einstein and Black Holes," presented at Harvard Black Hole Initiative, September 11, 2019.

Chapter 4: A Most Singular Solution

1. Brandon Carter, "Half Century of Black Hole Theory: From Physicists' Purgatory to Mathematicians' Paradise," arXiv:gr-qc/0604064, April 16, 2006.

2. Areeba Merriam, "Karl Schwarzschild's Letter to Albert Einstein," *Cantor's Paradise*, December 5, 2021, https://www.cantorsparadise.com/karl-schwarzschilds-letter-to-albert-einstein-6661734dd3e.

3. Karl Schwarzschild, "From Karl Schwarzschild," in Ann M. Hentschel, trans., *The Collected Papers of Albert Einstein*, vol. 8, *The Berlin Years: Correspondence, 1914–1918*, English translation supplement (Princeton, NJ: Princeton University Press, 1997), 163–165.

4. Ibid.

5. Albert Einstein, "To Karl Schwarzschild," in Ann M. Hentschel, trans., *The Collected Papers of Albert Einstein*, vol. 8, *The Berlin Years: Correspondence, 1914–1918*, English translation supplement (Princeton, NJ: Princeton University Press, 1997), 175–177.

6. Galina Weinstein, "Einstein, Schwarzschild, the Perihelion Motion of Mercury and the Rotating Disk Story," arXiv:1411.7370, November 26, 2014.

7. K. Schwarzschild, "On the Gravitational Field of a Sphere of Incompressible Fluid According to Einstein's Theory," trans. S. Antoci, arXiv:physics/9912033, December 16, 1999. (Originally published in *Sitzungsberichte der Königlich Preußischen Akademie der Wissenschaften zu Berlin [Math. Phys.]*, 1916, 424–434.)

8. Dennis Overbye, "A Century Ago, Einstein's Theory of Relativity Changed Everything," *New York Times*, November 24, 2015.

9. Arthur Eddington, "Relativistic Degeneracy," *The Observatory* 58:729 (1935), 37–39.

10. Marcia Bartusiak, *Black Hole: How an Idea Abandoned by Newtonians, Hated by Einstein, and Gambled on by Hawking Became Loved* (New Haven, CT: Yale University Press, 2015), 41.

11. Demetrios Christodoulou, "The Formation of Black Holes in General Relativity," arXiv:0806.3880, May 18, 2008, 5.

12. J. R. Oppenheimer and H. Snyder, "On Continued Gravitational Contraction," *Physical Review* 56:5 (September 1, 1939), 455–459.

13. Christodoulou, "The Formation of Black Holes in General Relativity," 5–6.

14. Albert Einstein, "On a Stationary System with Spherical Symmetry Consisting of Many Gravitating Masses," *Annals of Mathematics* 40 (October 1939), 922–936.

15. Petros S. Florides, "John Lighton Synge," *Biographical Memoirs of Fellows of the Royal Society* 54 (2018), 401–424.

16. Ibid.

17. Fulvio Melia, *Cracking the Einstein Code: Relativity and the Birth of Black Hole Physics* (Chicago: University of Chicago Press, 2009), 52–53.

18. Ibid., 70.

19. Roy Kerr, "Afterword," in Melia, *Cracking the Einstein Code*, 126–127.

20. Roy P. Kerr, "Gravitational Field of a Spinning Mass as an Example of Algebraically Special Metrics," *Physical Review Letters* 11:5 (1963), 237–238.

21. Melia, *Cracking the Einstein Code*, 1.

22. Kip Thorne, *Black Holes and Time Warps: Einstein's Outrageous Legacy* (New York: W. W. Norton, 1994), 290.

23. S. Chandrasekhar, "Shakespeare, Newton and Beethoven or Patterns of Creativity," *Current Science* 70 (May 1996), 810–822.

24. Florides, "John Lighton Synge."

25. Melia, *Cracking the Code*, 89.

26. Werner Israel, "Dark Stars: The Evolution of an Idea," in Stephen Hawking and Werner Israel, eds., *Three Hundred Years of Gravitation* (Cambridge, UK: Cambridge University Press, 1987), 253.

27. Roger Penrose, "Gravitational Collapse and Space-Time Singularities," *Physical Review Letters* 14 (January 18, 1965), 57–59.

28. Stephen Hawking, *A Brief History of Time* (New York: Bantam Books, 1988), 49.

29. Richard Schoen (Stanford University), interview with the author, January 31, 2008.

30. Thorne, *Black Holes and Time Warps*, 463.

31. Michael Brooks, "Cosmic Thoughts," *New Scientist* 256 (November 19, 2022), 46–49.
32. Ann Ewing, "'Black Holes' in Space," *Science News Letter*, January 18, 1964, 39. (The term *black holes* was used at the December 1963 annual meeting of the American Association for the Advancement of Science in Cleveland.)
33. Richard Schoen and S.-T. Yau, "The Existence of a Black Hole Due to Condensation of Matter," *Communications in Mathematical Physics* 90 (1983), 575–579.
34. S. W. Hawking, "Black Holes in General Relativity," *Communications in Mathematical Physics* 25 (1972), 152–166.
35. Gary T. Horowitz, "Higher Dimensional Generalizations of the Kerr Black Hole," arXiv:gr-qc/0507080, July 18, 2005.
36. Roberto Emparan and Harvey S. Reall, "A Rotating Black Ring Solution in Five Dimensions," *Physical Review Letters* 88:10–11 (March 2002), 101101.
37. Gregory J. Galloway and Richard Schoen, "A Generalization of Hawking's Black Hole Topology Theorem to Higher Dimensions," *Communications in Mathematical Physics* 266 (2006), 571–576.
38. Roger Penrose, "Gravitational Collapse: The Role of General Relativity," *Rivista del Nuovo Cimento* 1 (1969), 252–277.
39. Stephen W. Hawking, *The Theory of Everything* (Beverly Hills: Phoenix Books, 2005), 46.
40. Kevin Hartnett, "Mathematicians Disprove Conjecture Made to Save Black Holes," *Quanta*, May 17, 2018, https://www.quantamagazine.org/mathematicians-disprove-conjecture-made-to-save-black-holes-20180517/.
41. Ibid.
42. Jérémie Szeftel (Sorbonne University), interview with the author, June 25, 2021.
43. Sergiu Klainerman (Princeton University), email to the author, June 1, 2022.
44. Thibault Damour (IHÉS), email to the author, June 3, 2022.
45. Elena Giorgi, Sergiu Klainerman, and Jérémie Szeftel, "Wave Equations Estimates and the Nonlinear Stability of Slowly Rotating Kerr Black Holes," arXiv:2205.14808, May 31, 2022.
46. Elena Giorgi (Columbia University), interview with the author, June 24, 2022.
47. Robert Bartnik and John McKinnon, "Particlelike Solutions of the Einstein–Yang–Mills Equations," *Physical Review Letters* 61:2 (1988), 141–144.
48. Felix Finster (University of Regensburg), interview with the author, September 12, 2022.
49. Yuewen Chen, Jie Du, and Shing-Tung Yau, "Stable Black Hole with Yang–Mills Hair," arXiv:2210.03046, October 6, 2022.

50. The Royal Swedish Academy of Sciences, "The Nobel Prize in Physics 2020," press release, October 6, 2020, https://www.nobelprize.org/prizes/physics/2020/press-release/.

51. Brooks, "Cosmic Thoughts."

52. Lee Billings, "Black Hole Scientists Win Nobel Prize in Physics," *Scientific American*, October 6, 2020, https://www.scientificamerican.com/article/black-hole-scientists-win-nobel-prize-in-physics1/.

53. Stephen Hawking, "Foreword," in Thorne, *Black Holes and Time Warps*, 12.

Chapter 5: Chasing the Wave

1. Albert Einstein, "Approximate Integration of the Field Equations of Gravitation," in Ann M. Hentschel, trans., *The Collected Papers of Albert Einstein*, vol. 6, *The Berlin Years: Writings, 1914–1917*, English translation supplement (Princeton, NJ: Princeton University Press, 1997), 201–210. (Originally published on June 22, 1916.)

2. Abraham Pais, *Subtle Is the Lord: The Science and the Life of Albert Einstein* (New York: Oxford University Press, 2008), 22, 279.

3. Albert Einstein, "To Karl Schwarzschild," in Ann M. Hentschel, trans., *The Collected Papers of Albert Einstein*, vol. 8, *The Berlin Years: Correspondence, 1914–1918*, English translation supplement (Princeton, NJ: Princeton University Press, 1997), 196.

4. Albert Einstein, "Approximate Integration of the Field Equations of Gravitation," *Sitzungsberichte der Königlich Preußischen Akademie der Wissenschaften* (1916), 688–696.

5. Albert Einstein, "On Gravitational Waves," *Sitzungsberichte der Königlich Preußischen Akademie der Wissenschaften* (1918), 154–167.

6. Daniel Kennefick, "Controversies in the History of the Radiation Reaction Problem in General Relativity," arXiv:gr-qc/9704002, April 1, 1997.

7. Jorge L. Cervantes-Cota, Salvador Galindo-Uribarri, and George F. Smoot, "A Brief History of Gravitational Waves," arXiv:1609.09400, September 26, 2016.

8. Whitney Clavin, "When Black Holes Collide," *Caltech News*, January 24, 2019, https://www.caltech.edu/about/news/when-black-holes-collide-85110.

9. J. Hadamard, "*Sur les problèmes aux dérivées partielles et leur signification physique*," *Princeton University Bulletin* 13 (April 1902), 49–52.

10. Lydia Bieri, "Book Review: A Lady Mathematician in This Strange Universe: Memoirs," *Notices of the American Mathematical Society* 67:3 (March 2020), 387.

11. Lydia Bieri (University of Michigan), interview with the author, February 23, 2019.

12. Y. Choquet-Bruhat, "*Théorème d'existence pour certains systèmes d'équations aux dérivées partielles non linéaires*," *Acta Mathematica* 88:1 (1952), 141–225.

13. Bieri, "Book Review: A Lady Mathematician," 386.

14. Lydia Bieri, interview with the author, February 23, 2019.

15. Daniel Holz, "The Difficult Childhood of Gravitational Waves," *Discover*, April 25, 2007.

16. Frans Pretorius (Princeton University), interview with the author, September 8, 2021.

17. Bieri, "Book Review: A Lady Mathematician.'"

18. Martin Lesourd (Harvard Black Hole Initiative), interview with the author, December 13, 2019.

19. Yvonne Choquet-Bruhat and Robert Geroch, "Global Aspects of the Cauchy Problem in General Relativity," *Communications in Mathematical Physics* 14 (1969), 329–335.

20. Demetrios Christodoulou and Sergiu Klainerman, *The Global Nonlinear Stability of the Minkowski Space (PMS–41)* (Princeton, NJ: Princeton University Press, 1993).

21. Mihalis Dafermos (Princeton University), email to the author, April 6, 2020.

22. Ibid.

23. Demetrios Christodoulou, "Nonlinear Nature of Gravitation and Gravitational-Wave Experiments," *Physical Review Letters* 67:12 (1991), 1486–1489.

24. Lydia Bieri, Po-Ning Chen, and Shing-Tung Yau, "The Electromagnetic Christodoulou Memory Effect and Its Application to Neutron Star Binary Mergers," arXiv:1110.0410, October 3, 2011.

25. Paul D. Lasky, Eric Thrane, Yuri Levin, Jonathan Blackman, and Yanbei Chen, "Detecting Gravitational-Wave Memory with LIGO: Implications of GW150914," *Physical Review Letters* 117 (August 5, 2016), 061102.

26. Mihalis Dafermos, email to the author, April 6, 2020.

27. The Royal Swedish Academy of Sciences, "The Nobel Prize in Physics 1993," press release, October 13, 1993, https://www.nobelprize.org/prizes/physics/1993/press-release/.

28. Frans Pretorius, interview with the author, September 8, 2021.

29. Davide Castelvecchi, "What 50 Gravitational-Wave Events Reveal About the Universe," *Nature*, October 30, 2020, https://www.nature.com/articles/d41586-020-03047-0.

Chapter 6: An Equation for the Whole Universe

1. Tim Folger, "Einstein's Grand Quest for a Unified Theory," *Discover*, September 29, 2004, https://www.discovermagazine.com/the-sciences/einsteins-grand-quest-for-a-unified-theory.

2. Albert Einstein, "To Paul Ehrenfest," in Ann M. Hentschel, trans., *The Collected Papers of Albert Einstein*, vol. 8, *The Berlin Years: Correspondence, 1914–1918*, English translation supplement (Princeton, NJ: Princeton University Press, 1997), 282.

3. Albert Einstein, "To Willem de Sitter," in Ann M. Hentschel, trans., *The Collected Papers of Albert Einstein*, vol. 8, *The Berlin Years: Correspondence, 1914–1918*, English translation supplement (Princeton, NJ: Princeton University Press, 1997), 301–302.

4. Albert Einstein, "Cosmological Considerations in the General Theory of Relativity," in Ann M. Hentschel, trans., *The Collected Papers of Albert Einstein*, vol. 6, *The Berlin Years: Writings, 1914–1917*, English translation supplement (Princeton, NJ: Princeton University Press, 1997), 421–432.

5. Albert Einstein, *Relativity: The Special and the General Theory* (Princeton, NJ: Princeton University Press, 2015), 153.

6. Arthur Eddington, *The Expanding Universe*, Revised Edition (Cambridge, UK: Cambridge University Press, 1988), 21. (Originally published in 1933.)

7. Einstein, *Relativity*, 153.

8. Donald Goldsmith, *Einstein's Greatest Blunder? The Cosmological Constant and Other Fudge Factors in the Physics of the Universe* (Cambridge, MA: Harvard University Press, 1997).

9. Cormac O'Raifeartaigh and Simon Mitton, "Interrogating the Legend of Einstein's 'Biggest Blunder,'" arXiv:1804.06768, February 27, 2019.

10. Robbert Dijkgraaf, "Without Einstein, We'd All Be Lost," *Wall Street Journal*, November 5, 2015.

11. Einstein, "To Willem de Sitter," 308–309.

12. O'Raifeartaigh and Mitton, "Interrogating the Legend of Einstein's 'Biggest Blunder.'"

13. Abraham Pais, *Subtle Is the Lord: The Science and the Life of Albert Einstein* (New York: Oxford University Press, 2008), 288.

14. A. S. Eddington, *The Mathematical Theory of Relativity* (Cambridge, UK: Cambridge University Press, 1923), 272–273.

15. Eddington, *The Expanding Universe*, 46.

16. Alexander Friedmann, "On the Curvature of Space," *Zeitschrift für Physik* 10 (1922), 377–386.
17. Stephen Hawking, *A Brief History of Time* (New York: Bantam Books, 1988), 40.
18. Lisa Randall, "Energy in Einstein's Universe," in Peter L. Galison, Gerald Holton, and Silvan S. Schweber, eds., *Einstein for the 21st Century: His Legacy in Science, Art, and Modern Culture* (Princeton, NJ: Princeton University Press, 2008), 305.
19. Harry Nussbaumer and Lydia Bieri, *Discovering the Expanding Universe* (Cambridge, UK: Cambridge University Press, 2009), 90.
20. J. J. O'Connor and E. F. Robertson, "Aleksandr Aleksandrovich Friedmann," *MacTutor*, December 1997, https://mathshistory.st-andrews.ac.uk/Biographies /Friedmann/.
21. Martin Harwit, *In Search of the True Universe: The Tools, Shaping, and Cost of Cosmological Thought* (Cambridge, UK: Cambridge University Press, 2013), 42.
22. Tom Siegfried, "Einstein's Genius Changed Science's Perception of Gravity," *Science News*, October 4, 2015, https://www.sciencenews.org/article/einsteins -genius-changed-sciences-perception-gravity.
23. Abhay Ashtekar, "Geometry and Physics of Null Infinity," in Lydia Bieri and Shing-Tung Yau, eds., *One Hundred Years of General Relativity: A Jubilee Volume on General Relativity and Mathematics*, Surveys in Differential Geometry XX (Boston: International Press, 2015), 99.
24. Jean-Pierre Luminet, "Lemaître's Big Bang," arXiv:1503.08304, March 28, 2015.
25. Hawking, *A Brief History of Time*, 49–50.
26. S. W. Hawking and R. Penrose, "The Singularities of Gravitational Collapse and Cosmology," *Proceedings of the Royal Society A* 314 (1970), 529–548.

Chapter 7: The Matter of Mass (and the Mass of Matter)

1. R. Penrose, R. D. Sorkin, and E. Woolgar, "A Positive Mass Theorem Based on the Focusing and Retardation of Null Geodesics," arXiv:gr-qc/9301015, January 15, 1993.
2. Richard Schoen and Shing-Tung Yau, "On the Proof of the Positive Mass Conjecture in General Relativity," *Communications in Mathematical Physics* 65:1 (1979), 45–76.
3. Richard Schoen and Shing-Tung Yau, "Proof of the Positive Mass Theorem. II," *Communications in Mathematical Physics* 79 (1981), 231–260.

4. Hubert L. Bray, "Proof of the Riemannian Penrose Conjecture Using the Positive Mass Theorem," arXiv:math/9911173, November 23, 1999.

5. Edward Witten, "A New Proof of the Positive Energy Theorem," *Communications in Mathematical Physics* 80:3 (1981), 381–402.

6. Richard Schoen and Shing-Tung Yau, "Proof That the Bondi Mass Is Positive," *Physical Review Letters* 48:6 (February 8, 1982), 369–371.

7. Richard Schoen and Shing-Tung Yau, "Positive Scalar Curvature and Minimal Hypersurface Singularities," arXiv:1704.05490, April 18, 2017.

8. R. Penrose, "Some Unsolved Problems in Classical General Relativity," in Shing-Tung Yau, ed., *Seminar on Differential Geometry* (Princeton, NJ: Princeton University Press, 1982), 631–668.

9. Ibid., 631.

10. Ibid., 635.

11. Stephen Hawking, "Gravitational Radiation in an Expanding Universe," *Journal of Mathematical Physics* 9 (1968), 598–604.

12. Robert Bartnik, "New Definition of Quasilocal Mass," *Physical Review Letters* 62 (1989), 2346–2348.

13. Mu-Tao Wang (Columbia University), interview with the author, June 26, 2022.

14. Lan-Hsuan Huang (University of Connecticut), conversation with the author, April 29, 2022.

15. Mu-Tao Wang, interview with the author, January 15, 2022.

16. J. David Brown and James W. York, Jr., "Quasilocal Energy in General Relativity," *Contemporary Mathematics* 132 (1992), 129–142.

17. Mu-Tao Wang, email to the author, February 18, 2022.

18. Yuguang Shi and Luen-Fai Tam, "Positive Mass Theorem and the Boundary Behaviors of Compact Manifolds with Nonnegative Scalar Curvature," *Journal of Differential Geometry* 62 (2002), 79–125.

19. Mu-Tao Wang and Shing-Tung Yau, "Quasilocal Mass in General Relativity," arXiv:0804.1174, April 8, 2008.

20. Ibid.

21. Mu-Tao Wang, interview with the author, March 17, 2020.

22. Po-Ning Chen, Mu-Tao Wang, and Shing-Tung Yau, "Conserved Quantities in General Relativity: From the Quasi-Local Level to Spatial Infinity," *Communications in Mathematical Physics* 338 (2015), 31–80.

23. Po-Ning Chen, Mu-Tao Wang, Ye-Kai Wang, and Shing-Tung Yau, "Conserved Quantities in General Relativity—The View from Null Infinity," arxiv: 2204.04010, April 8, 2022.

24. This analogy was suggested by David Garfinkle during a conversation with the author on January 26, 2022.

25. Abhay Ashtekar (Pennsylvania State University), interview with the author, January 26, 2022.

26. Penrose, "Some Unsolved Problems in Classical General Relativity."

27. Mu-Tao Wang, interview with the author, January 21, 2022.

28. Demetrios Christodoulou, "Report on the Paper: Supertranslation Invariance of Angular Momentum," sent via email to the authors on November 6, 2021.

29. Lydia Bieri (University of Michigan), email to the author, December 3, 2021.

30. Vijay Varma (Albert Einstein Institute), interview with the author, June 27, 2022.

31. Lydia Bieri, email to the author, December 3, 2021.

Chapter 8: The Quest for Unification

1. Albert Einstein, "Fundamental Ideas and Problems of the Theory of Relativity," presented at the Nordic Assembly of Naturalists, Gothenburg, July 11, 1923. (Available at https://www.nobelprize.org/uploads/2018/06/einstein-lecture.pdf.)

2. Walter Isaacson, *Einstein: His Life and Universe* (New York: Simon & Schuster, 2007), 337.

3. David Gross, "Einstein and the Quest for a Unified Theory," in Peter L. Galison, Gerald Holton, and Silvan S. Schweber, eds., *Einstein for the 21st Century: His Legacy in Science, Art, and Modern Culture* (Princeton, NJ: Princeton University Press, 2008), 287.

4. "Einstein's Quest for a Unified Theory," *APS News* 14:11 (December 2005), https://www.aps.org/publications/apsnews/200512/history.cfm.

5. Abraham Pais, *Subtle Is the Lord: The Science and the Life of Albert Einstein* (New York: Oxford University Press, 2008), 350.

6. Ibid.

7. Gross, "Einstein and the Quest for a Unified Theory," 287, 298.

8. Albert Einstein, *The Albert Einstein Collection*, vol. 2, *Essays in Science, Letters to Solovine, and Letters on Wave Mechanics* (Philosophical Library/Open Road, 2019).

9. Pais, *Subtle Is the Lord*, 325.

10. Einstein, *The Albert Einstein Collection*, vol. 2.

11. Albert Einstein, "On the Method of Theoretical Physics," *Philosophy of Science* 1:2 (April 1934), 167.

12. Jürgen Neffe, *Einstein: A Biography* (New York: Farrar, Straus and Giroux, 2007), 356.

13. Hermann Weyl, "Gravitation and Electricity," *Sitzungsberichte der Königlich Preußischen Akademie der Wissenschaften zu Berlin* (1918), 465–480.

14. Lochlainn O'Raifeartaigh, *The Dawning of Gauge Theory* (Princeton, NJ: Princeton University Press, 1997), 45.

15. Juan Maldacena, "The Economic Analogy," *Plus*, July 16, 2016, https://plus.maths.org/content/its-economy-stupid.

16. Albert Einstein, "To Hermann Weyl," in Ann M. Hentschel, trans., *The Collected Papers of Albert Einstein*, vol. 8, *The Berlin Years: Correspondence, 1914–1918*, English translation supplement (Princeton, NJ: Princeton University Press, 1997), 654. (Letter originally dated September 27, 1918.)

17. Pais, *Subtle Is the Lord*, 341.

18. Michael Atiyah, *Hermann Weyl: 1885–1955*, Biographical Memoirs 82 (Washington, D.C.: The National Academy Press, 2002), 12.

19. Freeman J. Dyson, "Prof. Hermann Weyl, For.Mem.R.S.," *Nature* 177 (1956), 457–458.

20. Hermann Weyl, "Gravitation and the Electron," *Proceedings of the National Academy of Sciences* 15:4 (1929), 323–334.

21. Hermann Weyl, "*Elektron und Gravitation*," *Zeitschrift für Physik* 56 (1929), 330–352.

22. O'Raifeartaigh, *The Dawning of Gauge Theory*, vii.

23. Ibid.

24. Shiing-Shen Chern, "Geometry and Physics," presented at the University of Singapore, June 27, 1980.

25. O'Raifeartaigh, *The Dawning of Gauge Theory*, vii.

26. Atiyah, *Hermann Weyl*, 13.

27. Theodor Kaluza, "On the Unification Problem of Physics," in O'Raifeartaigh, *The Dawning of Gauge Theory*, 53–58. (Originally published in *Sitzungsberichte der Königlich Preußischen Akademie der Wissenschaften zu Berlin [Math. Phys.]* 96 [1921], 69–72.)

28. Oskar Klein, "*Quantentheorie und fünfdimensionale Relativitätstheorie*," *Zeitschrift für Physik* 37 (1926), 895–906; Oskar Klein, "The Atomicity of Electricity as a Quantum Theory Law," *Nature* 118 (1926), 516.

29. Albert Einstein, "To Theodor Kaluza," in Ann M. Hentschel, trans., *The Collected Papers of Albert Einstein*, vol. 9, *The Berlin Years: Correspondence, January*

1919–April 2020, English translation supplement (Princeton, NJ: Princeton University Press, 1997), 21. (Letter originally dated April 21, 1919.)

30. Brian Greene, *The Elegant Universe: Superstrings, Hidden Dimensions, and the Quest for the Ultimate Theory* (New York: Vintage Books, 1999), 203.

31. Eugenio Calabi (University of Pennsylvania), interview with the author, October 18, 2007.

32. James B. Hartle, "General Relativity in the Undergraduate Physics Curriculum," arXiv:gr-qc/0506075, February 3, 2008.

33. K. C. Cole, "From This Angle, Geometry Rules the Universe," *Los Angeles Times*, November 4, 1999, https://www.latimes.com/archives/la-xpm-1999-nov-04-me-30000-story.html.

34. Leonard Susskind, "Some Thoughts About String Theory and the World," Monday Colloquium at Harvard University Department of Physics, October 26, 2020.

35. Ibid.

36. Andrew Strominger and Cumrun Vafa, "Microscopic Origin of the Bekenstein–Hawking Entropy," *Physics Letters B* 379 (June 27, 1996), 99–104.

37. Brian Greene and Ronen Plesser, "Duality in Calabi–Yau Moduli Space," *Nuclear Physics B* 338 (1990), 15–37.

38. Philip Candelas, Xenia de la Ossa, Paul S. Green, and Linda Parkes, "A Pair of Calabi–Yau Manifolds as an Exactly Soluble Superconformal Theory," *Nuclear Physics B* 359 (1991), 21–74.

39. Andrew Strominger, Shing-Tung Yau, and Eric Zaslow, "Mirror Symmetry Is *T*-duality," *Nuclear Physics B* 479 (November 1996), 243–259.

40. Peter Galison (Harvard University), interview with the author, June 12, 2019.

41. Lewis Pyenson, "Einstein's Education: Mathematics and the Laws of Nature," *Isis* 71:3 (September 1980), 419.

42. Leo Corry, "The Influence of David Hilbert and Hermann Minkowski on Einstein's Views over the Interrelation Between Physics and Mathematics," *Endeavor* 22: 3 (1998), 95–97.

43. Constance Reid, *Hilbert* (London: Allen & Unwin, 1970), 127.

44. Chen Ning Yang, "Albert Einstein: Opportunity and Perception," *International Journal of Modern Physics A* 21 (2006), 3031–3038.

45. Mihalis Dafermos (Princeton University), interview with the author, April 2, 2020.

46. Hung-Hsi Wu (University of California, Berkeley), interview with the author, February 21, 2019.

47. Judith R. Goodstein, *Einstein's Italian Mathematicians: Ricci, Levi-Civita, and the Birth of General Relativity* (Providence, RI: American Mathematical Society, 2018), 145.

48. Peter Galison, "The Suppressed Drawing: Paul Dirac's Hidden Geometry," *Representations* 72 (Autumn 2000), 158.

49. Simon Donaldson (Imperial College), interview with the author, July 5, 2019.

50. Simon Donaldson, interview with the author, April 3, 2008.

51. Simon Donaldson, interview with the author, July 5, 2019.

52. Steve Mirsky, "A New Book Examines the Relationship Between Math and Physics," *Scientific American*, August 1, 2019, https://www.scientificamerican.com /article/a-new-book-examines-the-relationship-between-math-and-physics/.

Postlude

1. Nicholas Jackson, "St. Ignace Mystery Spot," *Atlas Obscura*, October 3, 2010, https://www.atlasobscura.com/places/st-ignace-mystery-spot.

2. "Mystery Spots," *RoadsideAmerica.com*, accessed September 14, 2023, https:// www.roadsideamerica.com/story/29062.

3. Tony Phillips, "NASA Announces Results of Epic Space-Time Experiment," *NASA Science*, May 4, 2011, https://einstein.stanford.edu/content/press-media /results_news_2011/NASA_ScienceNews.pdf.

4. Clifford M. Will, "Finally, Results from Gravity Probe B," *Physics*, May 31, 2011, https://physics.aps.org/articles/v4/43.

5. G. Voisin, I. Cognard, P. C. C. Freire, N. Wex, L. Guillemot, G. Desvignes, M. Kramer, and G. Theureau, "An Improved Test of the Strong Equivalence Principle with the Pulsar in a Triple Star System," *Astronomy and Astrophysics* 638 (June 2020), A24.

6. Max Planck Society, "Confirming Einstein's Most Fortunate Thought," press release, June 10, 2020, https://www.mpg.de/14923530/general-relativity -pulsar.

7. R. Abuter, A. Amorim, M. Bauböck, J. P. Berger, et al., "Detection of the Schwarzschild Precession in the Orbit of the Star S2 near the Galactic Centre Massive Black Hole," *Astronomy and Astrophysics* 636 (April 2020), L5.

8. Gabriella Agazie, Akash Anumarlapudi, Anne M. Archibald, Zaven Arzoumanian, et al., "The NANOGrav 15 yr Data Set: Evidence for a Gravitational-Wave Background," *The Astrophysical Journal Letters* 951:1 (2023), L8.

9. Katrina Miller, "The Cosmos Is 'Thrumming' with Gravitational Waves, Astronomers Find," *New York Times*, June 28, 2023, https://www.nytimes.com/2023/06/28/science/astronomy-gravitational-waves-nanograv.html.

10. E. K. Anderson, C. J. Baker, W. Bertsche, N. M. Bhatt, et al., "Observation of the Effect of Gravity on the Motion of Antimatter," *Nature* 621 (September 28, 2023), 716–722.

11. Helmut Friedrich, "On the Existence of n-Geodesically Complete or Future Complete Solutions of Einstein's Field Equations with Smooth Asymptotic Structure," *Communications in Mathematical Physics* 107 (1986), 587–609.

12. Demetrios Christodoulou and Sergiu Klainerman, *The Global Nonlinear Stability of the Minkowski Space (PMS–41)* (Princeton, NJ: Princeton University Press, 1993).

13. Georgios Moschidis, "A Proof of the Instability of the AdS for the Einstein–Null Dust System with an Inner Mirror," arXiv:1704.08681, April 27, 2017.

14. Lars Andersson, Pieter Blue, Zoe Wyatt, and Shing-Tung Yau, "Global Stability of Spacetimes with Supersymmetric Compactifications," arXiv:2006.00824, June 1, 2020.

15. Marcus A. Khuri and Jordan F. Rainone, "Black Lenses in Kaluza–Klein Matter," arXiv:2212.06762v2, July 11, 2023.

16. Sven Hirsch, Demetre Kazaras, Marcus Khuri, and Yiyue Zhang, "Spectral Torical Band Inequalities and Generalizations of the Schoen–Yau Black Hole Existence Theorem," *International Mathematics Research Notices* (June 26, 2023), rnad129.

Index

Steve Nadis, a graduate of Hampshire College, is a contributing editor at *Discover* magazine and a contributing writer to *Quanta*. He lives in Cambridge, Massachusetts.

Shing-Tung Yau is a professor of mathematics at Tsinghua University and professor emeritus at Harvard University. The recipient of the Fields Medal, the National Medal of Science, a MacArthur Fellowship, and other awards, he lives in Beijing.

This is the fifth book that Nadis and Yau have written together.